Aspiration Cytology of the Breast

Dr. Hubert Schöndorf

Department of Obstetrics and Gynecology,
Johann Wolfgang Goethe-Universitat
Frankfurt/Main, Germany

Translated by

Volker Schneider, M.D.

Department of Pathology,
Montefiore Hospital and Medical Center
Bronx, New York

1978

W. B. SAUNDERS COMPANY / Philadelphia / London / Toronto

W. B. Saunders Company: West Washington Square
Philadelphia, PA   19105

1 St. Anne's Road
Eastbourne, East Sussex BN21 3UN, England

1 Goldthorne Avenue
Toronto M8Z 5T9, Ontario, Canada

Authorized English Edition.   All rights reserved.
Original German Edition: Die Aspirationszytologie der Brustdrüse

© 1977 by F. K. Schattauer Verlag, Stuttgart
English Edition published 1978 by W. B. Saunders Company, Philadelphia, London, Toronto

Aspiration Cytology of the Breast   ISBN   0-7216-8013-5

Last digit is the print number: 9   8   7   6   5   4   3   2   1

Preface

In recent years, fine needle aspiration of the breast has gained increasing acceptance as a diagnostic procedure. The results obtained indicate that it may improve accuracy in the diagnosis of breast lesions, especially if used in combination with mammography. Furthermore, surgical exploration becomes unnecessary in a considerable number of patients. Information about the nature of the lesion thus is available to the surgeon as well as to the patient, and the presence of malignancy can be established well in advance of major surgery.

The object of this book is to provide an introduction to this procedure, its technical performance, and its appropriate applications, for the clinician (surgeon, radiologist, or gynecologist) who wants to become familiar with fine needle aspiration. For the cytopathologist and cytotechnician, this book is intended to be of help in the interpretation of smears obtained by needle aspiration.

Benign and malignant lesions of the breast as well as recurrent and metastatic disease are the principal subjects of discussion. A series of photomicrographs following each section illustrates the characteristic features of the lesions presented.

My special thanks go to Dr. Schmidt-Matthiesen, Professor of Gynecology and Chairman, Department of Gynecology and Obstetrics, The University Hospital in Frankfurt, for his constant encouragement and advice, particularly in areas of clinical relevance. I enjoyed the close cooperation I had with Dr. Naujoks, Professor of Gynecology and Head, Laboratory for Clinical Cytology, The University Hospital in Frankfurt, who offered valuable advice in the preparation of the manuscript. Ms. Doris Esser was unfailing in her support, especially in the days when fine needle aspiration was introduced into our department. I am indebted to Dr. Glätzner, Professor of Radiology, and to Dr. Leppien, Histopathologist in our institution. They both provided results that were essential to the development of fine needle aspiration techniques in this laboratory. I also want to thank all those physicians who referred their patients to our clinic, thus contributing invaluably to the collection of material presented here. Finally, my appreciation goes to F. K. Schattauer Verlag and to Professor Dr. Matis for their close cooperation and generous support in the preparation of this book.

H. Schöndorf

Contents

7

8

9

10

11

Aspiration Cytology of the Breast

1. Introduction

Inspection, palpation, and surgical exploration are traditional and well-established diagnostic procedures for the evaluation of breast lesions. However, with the addition of new radiological techniques to the diagnostic armamentarium, i.e., mammography, thermography, and xerography, a reevaluation of traditional methods became necessary. Scarring and fibrosis of the connective tissue following open biopsy cause alterations in the consistency and optical density of the breast. Because continuous radiological follow-up is compromised by surgical intervention, limitation of the number of surgical explorations is desirable. Nevertheless, a tissue sample is still considered essential before any major surgery is performed.

Under these circumstances, fine needle aspiration may fill a gap in the diagnostic evaluation of the breast. This technique provides a representative tissue sample for microscopic examination without interfering with the radiological appearance of the breast. However, needle aspiration is limited to the exploration of palpable lesions.

It is my intention to present the data and experience gained from using this technique in our institution over a period of five years. These data may contribute to a better understanding of the role of fine needle aspiration in the diagnosis of lesions of the breast.

2. Historical and regional development

Needle aspiration of cellular material for diagnostic purposes is a well-established procedure. In 1912, Ward[111] suggested the aspiration of enlarged lymph nodes in the evaluation of lymphoma. In 1921, Guthrie[49] published his experience with the aspiration of lymph nodes in Hodgkin's disease.

At Memorial Hospital in New York, Martin and Ellis[73] began to aspirate routinely all palpable tumors in 1926. After four years, they reported their experience with the aspiration of 1400 malignant lesions.[74] Represented in their series were lymph nodes and neoplasms of the nasopharynx as well as tumors of bone, lung, prostate, thyroid, and breast. Simplicity, rapid diagnosis, little inconvenience for patient and physician, and minimal discomfort for the patient were listed as major advantages of the procedure.

In 1952, Saphir[91] recommended fine needle aspiration for lesions of the breast occurring in pregnancy and during lactation as well as for the rapid differential diagnosis of carcinoma of the breast and such benign conditions as enlarging fibroadenoma, abscess, cyst, or mastitis. In 1956, Godwin[42] suggested fine needle aspiration for palpable lesions situated in deeper aspects of the breast and for tumors that were difficult to explore surgically.

Around 1950, several European clinicians began to express interest in fine needle aspiration. In France, Fogher,[32] Cornillot and Verhaeghe in Lille,[20] Castelain and Castelain[18] and Zajdela[116] in Paris, Bonneau and De The in Marseille,[10] and Marsan and Bertini[72] introduced fine needle aspiration as a diagnostic procedure. Prevention of wound-healing complications in areas of preoperative radiation was reported to be a particular advantage. Multiple aspirations were performed in elderly patients with advanced disease in order to assess the extent of the lesion non-invasively. All palpable lesions of the breast have been aspirated by Zajdela[116] since 1954. Results based on the so-called triple diagnosis, i.e., the combined clinical, radiological, and cytological examination of the breast, were reported by Verhaeghe and Cornillot[108] in 1969. In their series, the rate of false negative results dropped to 1 per cent with the simultaneous application of all three procedures. Kreuzer, Boquoi, and Meyer[65] were able to confirm these results.

Fine needle aspiration has been used extensively by Söderström in Lund[98] and by Franzen and Zajicek[34] at the Radiumhemmet in Stockholm. Franzen and Zajicek provided statistical data for this technique in a series of papers based on the largest sample ever reported. The propagation of needle aspiration in Germany was markedly advanced by these investigators. Franzen and Zajicek routinely aspirate all palpable lesions of the breast.

Reports on fine needle aspiration were published in England by Gibson and Smith,[40] Tribe,[106] and later by Webb[112] and Furnival and his associates.[35] Like some other investigators, Webb aspirates all palpable breast lesions. If an unequivocal cytological diagnosis of malignancy has been established preoperatively, Webb considers intraoperative examination of frozen sections unnecessary. Furthermore, in rapidly growing large carcinomas of the breast, open biopsy is more likely to produce tumor cell dissemination than is fine needle aspiration. In view of overcrowded clinics, the rapidity of the procedure is considered another advantage.

In Germany, fine needle aspiration of the breast is used mainly in gynecological departments[3, 11, 13, 27, 29, 94] for initial evaluation of palpable lesions as well as for postoperative follow-up. The prevailing opinion is that this procedure can improve diagnostic accuracy when used in combination with clinical examination, mammography, and thermography. Fine needle aspiration is being used increasingly by radiologists in Germany. Its application in fields of radiation is regarded as particularly helpful in the prevention of ulceration and tissue breakdown.[29] In addition, the procedure compensates for a number of shortcomings of mammography, as in peripheral carcinoma or medullary carcinoma. Thus, in the series published by Evers and Fischedick,[27] nine carcinomas of the breast were detected by cytological techniques alone, since clinical examination and mammography both gave negative results in these patients. We had a similar experience in our series.

Fine needle aspiration is performed in a number of institutions in the United States.[9, 19, 53, 90, 96] However, the procedure is far from being generally accepted. Fine needle aspiration is routinely performed in Memorial Hospital in New York,[53] where the method originated.

Fine needle aspiration of non-palpable lesions detected by mammography also has been reported.[11, 55] The targets were microcalcifications and indurations with formation of spicules. Despite additional instruments, the detection of such lesions by fine needle aspiration remains problematical. Therefore, this approach does not have practical significance at this time and therefore is not recommended.

3. Technique of fine needle aspiration

The requirements for a successful application of fine needle aspiration are: correct position of the needle, aspiration of sufficient material, and correct technical preparation of the smear (Figs. 1 and 2).

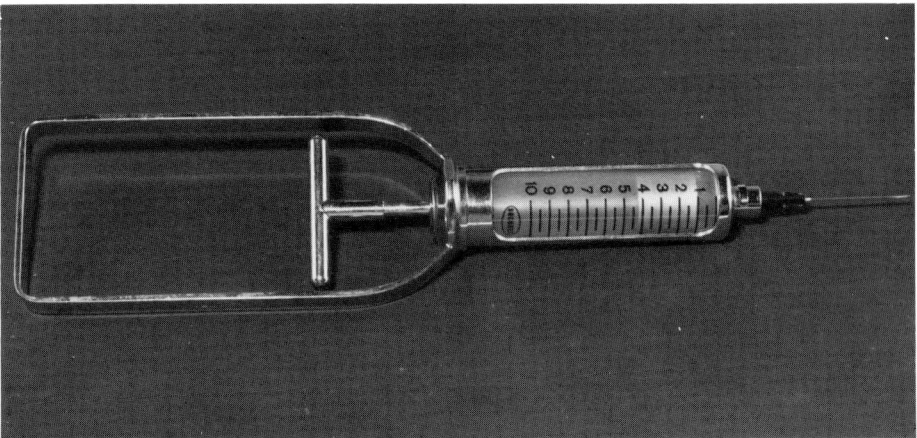

Figure 1. Syringe holder especially designed for fine needle aspiration. (Franzen)

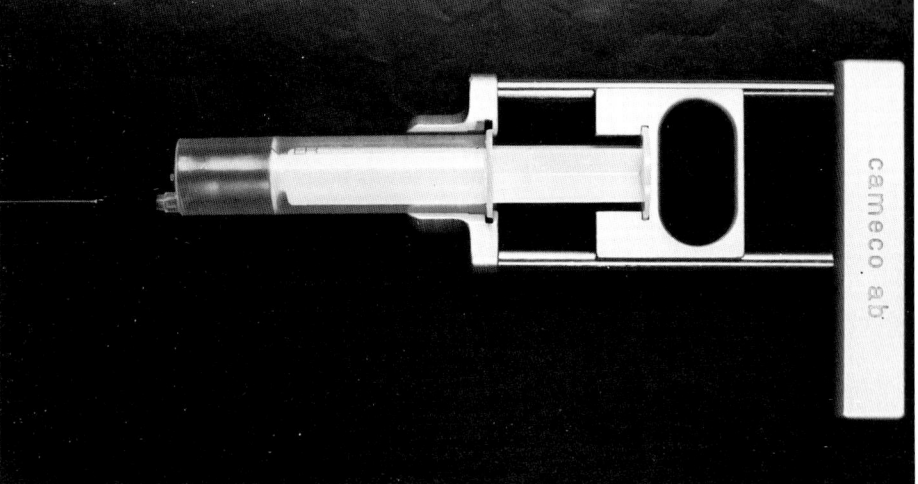

Figure 2. Special handle for fine needle aspiration, permitting the use of disposable syringes. (Cameco Company, Enebyberg, Sweden)

It is not necessary to use local anesthesia for needle aspiration, since this procedure is no more painful than an intramuscular injection. In a rare case of aspiration in the more sensitive area of the nipple, a topical anesthetic may be used. The skin overlying the lesion is first disinfected; surgical draping is not necessary. The use of a special syringe holder, designed by Franzen (Fig. 1), or a handle for disposable syringes, available from Cameco Company* (Fig. 2), permits a single-handed performance of the aspiration. Thus, the lesion can be immobilized with the index and middle fingers of the other hand (Fig. 3a and 3b).

After the needle has entered the lesion, the piston of the syringe is retracted, creating a negative pressure in the needle. Backward and forward movements are then performed with the needle tip remaining in the lesion, so that the area sampled corresponds roughly to a cone (Fig. 4). Before the needle is removed, the piston must be released to prevent aspiration of the material into the syringe.

---

*Cameco Company, Enebyberg, Sweden.

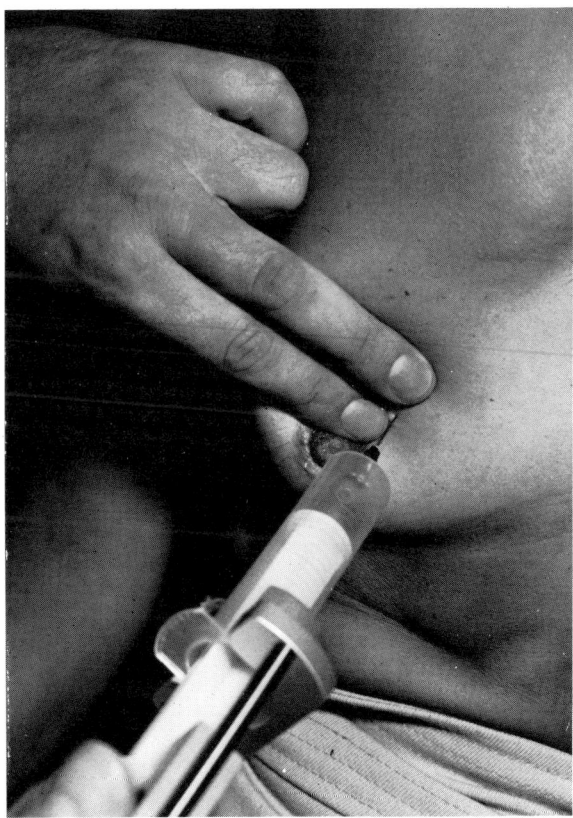

Figure 3a. Technique of fine needle aspiration of the breast.

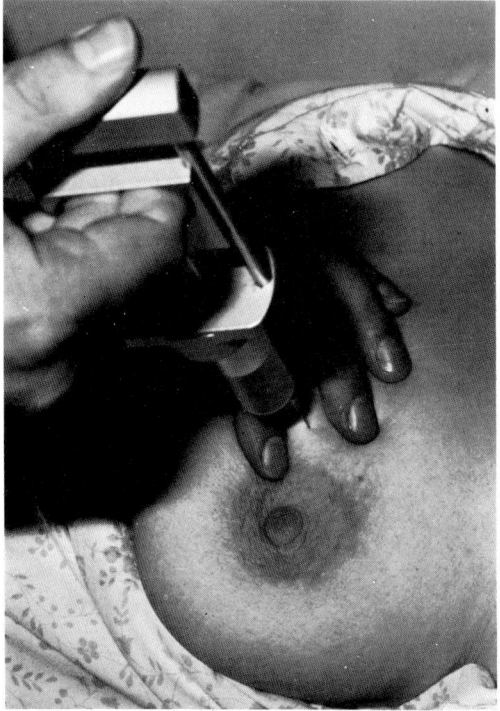

Figure 3b.   Technique of fine needle aspiration of the breast.

The needle is then withdrawn and detached, and the syringe is filled with air. The needle is reattached and the material in the needle expressed onto a slide (Fig. 6).

In 4 to 7 per cent of cases the needle does not contain sufficient material because of marked fibrosis of the lesion.[103] In these instances, a large bore needle may give better results.[118] However, hematomas do occur more often when larger needles are used. Multiple aspirations at various angles permit sampling of material from different areas of the lesion (Fig. 4a and 4b).[22] Generally, multiple aspirations are recommended in order to obtain sufficient material. In 25 out of 120 malignant lesions in this series, the final diagnosis was based on material obtained in the second or third aspiration. The center of a malignant tumor may occasionally undergo necrosis.[9] Aspiration of marginal areas of the lesion is recommended in this case.

There are a number of reasons why unsatisfactory specimens are obtained:

1. *Aspiration without a syringe holder.* Without a syringe holder, a single-handed aspiration cannot be done. Insufficient or non-representative specimens are obtained, since the lesion cannot be immobilized with the free hand.

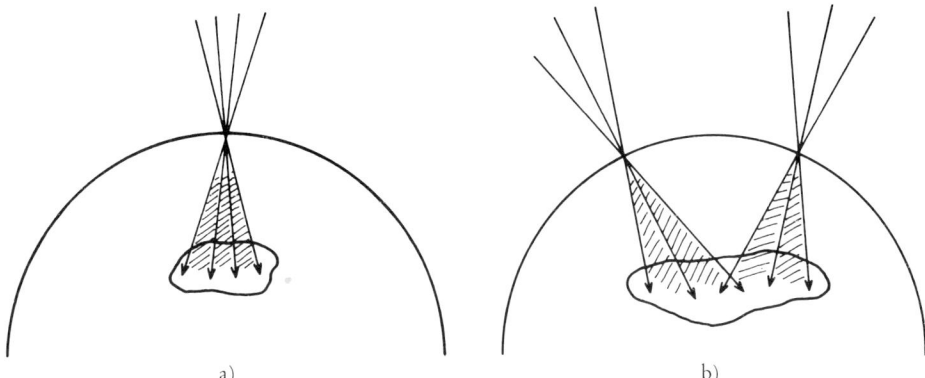

a)   b)

Figure 4. Schematic demonstration of the sampling technique in fine needle aspiration. a. Single aspiration. b. Multiple aspirations.

2. *Aspiration without moving the needle.* With the exception of cystic lesions, little cellular material is aspirated if the needle tip remains in the same position during the application of negative pressure.

3. *Removal of the needle before releasing the negative pressure.* If the piston of the syringe is not released before the needle is withdrawn, the material is aspirated into the syringe, from where it usually cannot be recovered.

4. *Small size of the lesion, i.e., less than 0.5 cm in diameter.* In this instance, the lesion probably will be missed. This is also true of non-palpable lesions detected by mammography.

5. *Fibrosis of the tumor.* The aspiration of densely fibrotic lesions is often unsatisfactory, owing to poor cellularity. Therefore, needle aspiration has a rather high failure rate in the diagnosis of scirrhous carcinoma.

4. Complications

In 30 per cent of all needle aspirations of the breast, a small hematoma is produced at the site of aspiration. Histologically, the bleeding occurs not in the lesion itself but in adipose tissue located between the skin and the tumor.[10] Local pressure with a surgical sponge prevents this minor complication. Mastitis or pneumothorax caused by needle aspiration was not seen in our series, but either may occur occasionally.[17, 34, 42, 65, 72, 73, 115] Evers and his colleagues[27] reported the inflammation of a cyst after pneumocystography.

Dissemination of tumor cells by needle aspiration is often cited as a serious complication. Since mastectomy usually follows immediately after aspiration of a malignant lesion, the formation of implantation metastases in the needle track has no practical relevance. However, the possibility that needle aspiration may induce distant metastatic spread via lymphatics or blood vessels must be considered.

Experimental data on this subject were provided by Engzell and his colleagues[25] in an investigation on rabbits. Efferent lymphatics and veins draining a popliteal metastasis of a carcinoma were catheterized. Lymphatic fluid and venous blood were collected and examined for the presence of carcinoma cells. After gentle massage of the tumor, carcinoma cells were found in the lymphatic fluid in one out of seven animals and in the venous blood in one out of nine animals. After needle aspiration of the metastasis, carcinoma cells could not be demonstrated either in lymphatic fluid or in venous blood.

Statistical data on the influence of needle aspiration in the survival of breast cancer patients were reported from Memorial Hospital in New York.[7, 87] Two groups, each consisting of 370 patients treated by mastectomy, were compared.

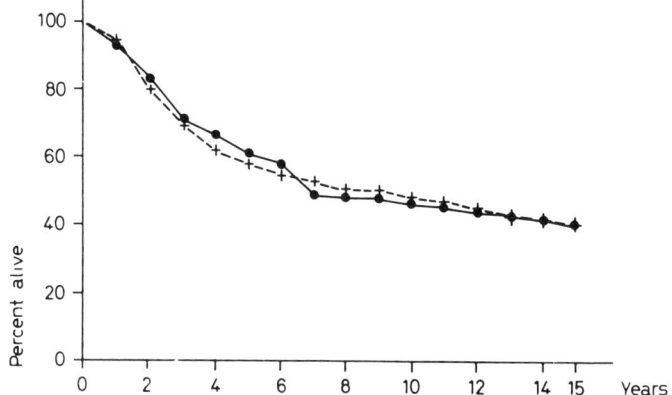

Figure 5. Fifteen-year actuarial survival rates for breast cancer patients who had aspiration biopsy and their matching controls. (*From* Berg, J. W., and Robbins, G. F. *In:* Cancer, 15:826, 1962)

In the first group, fine needle aspiration had been performed prior to mastectomy. The second group, which had not received needle aspiration, served as a matched control. The 15-year survival rate was found to be identical in both groups (Fig. 5).

Thus, experimental as well as statistical data indicate that dissemination of tumor cells by needle aspiration does not represent a hazard for the patient, particularly if this procedure is compared with the much more aggressive approach of the surgical biopsy.

5. Technical aspects

5.1. Preparation of the smear

A drop of the aspirate is placed on a slide and carefully spread with a coverslip. Cells located at the margins of the slide may be distorted if too much pressure is applied. Regardless of its color, the fluid obtained from the aspiration of cysts is always examined cytologically. Since cyst fluid contains few if any cells, centrifugation is necessary. In our laboratory, a cytocentrifuge (Shandon-Elliott*) is used (Figs. 46 and 47), and 0.4 ml of conventionally centrifuged fluid is submitted. The cellular elements are concentrated in an area of 28 mm^2 on the slide. Centrifugation time is 10 minutes at 1300 rounds per minute.

5.2. Fixation

Depending on the staining technique to be used, the smear should either be air-dried or fixed in alcohol. Slides to be stained with the technique of May-Grünwald-Giemsa should be air-dried. Air-dried smears may be held for several days or sent by mail without altering their preservation or staining quality. If the Papanicolaou technique is used, the immediate wet fixation of the smear is extremely important. Fixatives used are equal parts of 95 per cent ethyl alcohol and ether, isopropyl alcohol, or commercial aerosol spray fixatives. Non-viscous material suspended in small amounts of fluid may be partially lost owing to lack of adhesion to the slide surface. Adipose tissue also may be dissolved and lost in alcoholic fixatives (Fig. 6).

5.3. Staining procedures

The techniques of May-Grünwald-Giemsa and Papanicolaou are the ones most commonly used for needle aspirates. In our laboratory, like others,[37, 110] the May-Grünwald-Giemsa stain is preferred because of its simplicity and rapidity. The entire procedure does not take more than 30 minutes. Cytoplasm and nucleus both stain intensely with May-Grünwald-Giemsa. Background material such as mucus, secretions, erythrocytes, and polymorphonuclear leukocytes can be evaluated easily.[116] Nuclei appear slightly larger than in Papanicolaou-stained material (Factor 1.2), as shown by karyometric measurements in our material.

---

*Shandon Southern Instruments, Inc., Sewickley, Pennsylvania 15143.

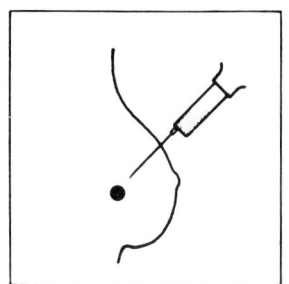

1. Aspiration

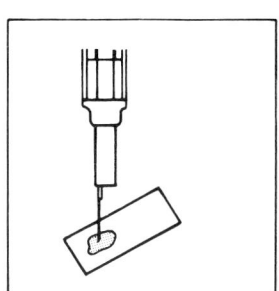

2. Ejection of
   the aspirate
   onto a slide

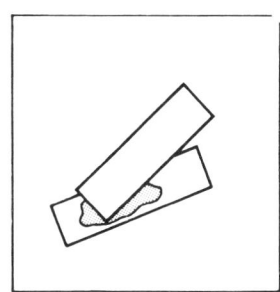

3. Preparation
   of the smear

4a. Air drying for
    May-Grünwald-
    Giemsa stain

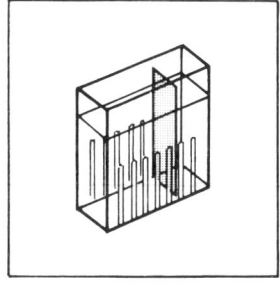

4b. Fixation for
    Papanico-
    laou stain

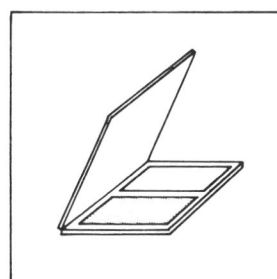

5. Ready to
   mail

Figure 6.   Processing of the aspirate.

Some laboratories prefer the Papanicolaou technique, which allows a more detailed evaluation of nuclear structure.[11, 83] Zajicek[122] recommends applying both stains concurrently, a procedure that is followed in our laboratory.

Solutions for May-Grunwald-Giemsa stain

Working solution No. 1 (May-Grünwald):
2 volumes of May-Grünwald reagent
1 volume of methyl alcohol
Working solution No. 2 (Giemsa):
1 volume of Giemsa reagent
9 volumes of distilled water

Both solutions should be prepared prior to use.

Staining procedure (air-dried smears)

1. May-Grünwald solution.	5 min
2. Rinse in running tap water without agitating.	30 sec
3. Giemsa solution.	15 min
4. Rinse in running tap water.	30 sec
5. Allow to air-dry	
6. Mount.	

Poor staining results may be due to the following:
1. Fixation in alcohol
2. Over- or understaining
3. Outdated reagents
4. Old Giemsa solution

Papanicolaou staining technique

1. 80 per cent ethyl alcohol	30 sec
2. 70 per cent ethyl alcohol	30 sec
3. 50 per cent ethyl alcohol	30 sec
4. Distilled water	30 sec
5. Harris' hematoxylin	3 to 6 min
6. Distilled water	30 sec
7. 0.25 per cent aqueous solution of hydrochloric acid	6 dips
8. Running tap water	6 min
9. Distilled water	30 sec
10. 50 per cent ethyl alcohol	30 sec
11. 70 per cent ethyl alcohol	30 sec
12. 80 per cent ethyl alcohol	30 sec
13. 95 per cent ethyl alcohol	30 sec
14. Orange G 6	30 sec
15. 95 per cent ethyl alcohol	30 sec

16. 95 per cent ethyl alcohol	30 sec
17. EA 50	$1^{1}/_{2}$ min
18. 95 per cent ethyl alcohol	30 sec
19. 95 per cent ethyl alcohol	30 sec
20. 95 per cent ethyl alcohol	30 sec
21. Absolute alcohol	30 sec
22. Absolute alcohol and xylene	30 sec
23. Xylene	30 sec

The following modified Papanicolaou technique is used in our laboratory. Minimal fixation time in ether alcohol or isopropyl alcohol is 20 to 30 minutes. To prevent carryover, let excess solution drip off on a piece of paper towel after each step.

1. Tap water	2 min
2. Harris' hematoxylin	2 min
3. Tap water	3 min
4. Tap water	3 min
5. Orange G 6	3 min
6. Tap water	3 min
7. Tap water	3 min
8. Polychrome EA 31	2 min
9. Ethyl alcohol, 70 per cent	5 min
10. Ethyl alcohol, 96 per cent	5 min
11. Xylene, isopropyl alcohol $\overline{aa}$	5 min
12. Xylene	5 min

According to Boschann,[12] poor staining results are due to the following:

1. Air-drying effect
2. Improper fixative
3. Dirty or fatty slides
4. Contaminated alcohol solution
5. Outdated reagents
6. Insufficient rinsing
7. Over- or understaining
8. Insufficient clearing

6. Cellular elements in aspirates from the breast

6.1. Anatomy of the mammary gland

In the sexually mature woman, the mammary gland consists of 15 to 20 radially arranged lobes, which are separated by dense fibrous tissue. Each lobe contains multiple lobules, the lobule being the structural unit of the breast. Lobules are composed of secretory acini, terminal ducts, and intralobular connective tissue. The drainage system of the breast provides each lobe with a separate lactiferous duct that dilates in the area of the nipple to form the lactiferous sinus.

Acini and terminal ducts are lined with simple columnar or cuboidal epithelium. A second layer of cells, the so-called myoepithelial cells, may occasionally be visualized between the duct epithelium and the basement membrane. Myoepithelial cells have a spindly appearance, owing to multiple cytoplasmic processes. Major ducts are lined with two layers of cuboidal epithelium. Myoepithelial cells are rarely observed in the distal parts of the excretory system. In the area of the nipples, in their most distal portions, lactiferous ducts are lined with nonkeratinizing squamous epithelium. In contrast to the dense and poorly cellular interlobular stroma, the intralobular stroma and periductal connective tissue are characterized by loose texture with small amounts of collagen, rich vascularization, numerous elastic fibers, and absence of adipose tissue.

6.2. Cellular elements in benign conditions

Before aspirating the palpated lesion, the physician passes the needle through all the normal structures of the mammary gland, e.g., lobules, terminal ducts, major ducts, and connective septae. Therefore, elements of normal breast tissue are regularly observed in specimens obtained by needle aspiration. These are: epithelial cells deriving from ducts and acini, naked bipolar nuclei, apocrine cells (oncocytes), foam cells, fat cells, and fibrocytes.

Additional cellular elements that may be present are erythrocytes, lymphocytes, leukocytes, histiocytes, and giant cells. Amorphous material, such as secretions, necrotic debris, or mucus, occasionally may be encountered (Fig. 7).

6.2.1. Duct cells and acinar cells

Epithelial cells deriving from ducts and lobules are characterized by round or oval regularly shaped nuclei with dense chromatin. A scant rim of cytoplasm is usually present. When they occur in clusters, the cells present in a honey-

16

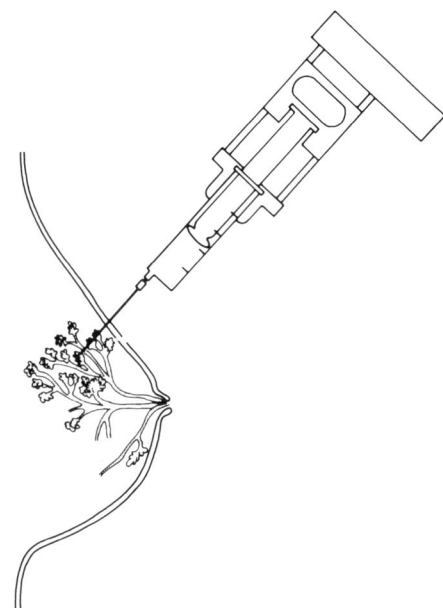

Figure 7. Normal tissue components of the mammary gland that are penetrated by the needle during aspiration.

comb pattern (Fig. 10). Smaller groups of epithelial cells present occasionally in a tubular or lobular arrangement (Figs. 48 and 49). In singly occurring epithelial cells, the cytoplasm is often absent. In comparison with naked bipolar nuclei, the nuclei of epithelial cells retain their round appearance and are slightly larger. Nucleoli are commonly observed in Papanicolaou-stained specimens. In aspirates from fibroadenoma and fibrocystic disease, nuclei with marked pleomorphism may occur, occasionally reaching the size of nuclei of malignant cells.[61, 72, 85] In these instances (Figs. 51, 52, and 53), it is particularly important to evaluate the entire cell population. In May-Grünwald-Giemsa-stained material, epithelial cells have a nuclear diameter of 8.4 $\pm$ 2 μ and appear slightly larger than in Papanicolaou-stained specimens (Factor 1.2), as shown by karyometric measurements in our material.

6.2.2. Naked bipolar nuclei

Bipolar nuclei having a characteristic oval or bipolar configuration are regularly observed in aspirates of the breast (Figs. 12, 13, and 14). Bipolar nuclei always occur singly and are slightly smaller than duct cell nuclei, measuring on the average 6 to 8 μ. The chromatin pattern is either finely granular and homogeneous or pyknotic and hyperchromatic. According to Murad,[83] bipolar nuclei may occur as light or dark forms. The origin of bipolar nuclei is not yet definitely established. Some authors believe they derive from myoepithelial cells.[11, 122]

Compared with duct cells, bipolar nuclei have a much more uniform appearance and show less variation in size and shape.

6.2.3. Apocrine cells (Oncocytes)

Apocrine cells are a common finding in needle aspirates of the breast. They are often present in small cysts, forming papillary projections. In aspirates of large cysts, they are sometimes the only cellular elements encountered. Their cell borders are distinct and prominent; the nuclei are round, clear, and usually centrally located (Figs. 15 and 16). The cytoplasm contains numerous eosinophilic granules (Fig. 17) that have been shown to correspond ultrastructurally to swollen mitochondria. Whereas nuclear size may vary considerably in diameter, from 6 to 11 μ, nuclear shape is rather uniform. The cytoplasm usually stains cyanophilic; in Papanicolaou-stained smears, it occasionally appears eosinophilic.

6.2.4. Foam cells

As indicated by their name, these cells are characterized by a fine vacuolization of the cytoplasm, which gives them a foamy appearance. Foam cells show marked variation in size. The nuclei, which are usually in a marginal position, are round, with a prominent nuclear membrane and finely granular chromatin. Multinucleation may occur (Figs. 18 and 46).

The precise origin of foam cells is unknown. They are believed to derive either from epithelial cells[58, 85] or from histiocytes.[85, 91] Their capacity for phagocytosis and their morphological similarity to histiocytes favor a histiocytic origin. On the other hand, a continuous spectrum of cells, representing all transitional forms from duct cells to foam cells, is often observed and makes the histiocytic origin of foam cells rather unlikely.

6.2.5. Fat cells

Fat cells usually occur in clusters. They contain small, dark nuclei in a peripheral position. The cell borders are prominent, with an exceedingly thin rim of cytoplasm (Fig. 19).

6.2.6. Fibrocytes

Fibrocytes are elements of the connective tissue. Occasionally, rather large fragments of stromal tissue are present in needle aspirates of the breast. Fibrocytes may be easily recognized at the margins of these tissue fragments. They have a spindly appearance and a centrally located nucleus that is round, oval, or spindly. In general, fibrocytes are encountered only rarely in needle aspirates (Figs. 20 and 21).

6.2.7.   Giant cells

The formation of giant cells is not limited to a single cell type. Rather, the presence of multiple nuclei is a common finding in a variety of conditions. In the material we studied, giant cells were found in pregnancy (Fig. 25), in the early postpartum-period,[99] in inflammatory reactions (Fig. 22), in granulation tissue as typical foreign body-type giant cells (Fig. 24), and occasionally in cyst fluid (Fig. 23). Vassilakos[107] reported the aspiration of a tuberculous granuloma of the breast containing typical Langhans-type giant cells. In carcinoma of the breast, numerous multinucleated giant cells may be found representing atypical tumor cells (Fig. 38). Giant cells with extremely bizarre nuclei may be observed after radiation therapy (Fig. 39).

Figures 8 through 25 appear in the following section.

Figure 8.   Aspiration of the breast. Normal epithelial cells. May-Grünwald-Giemsa × 400.

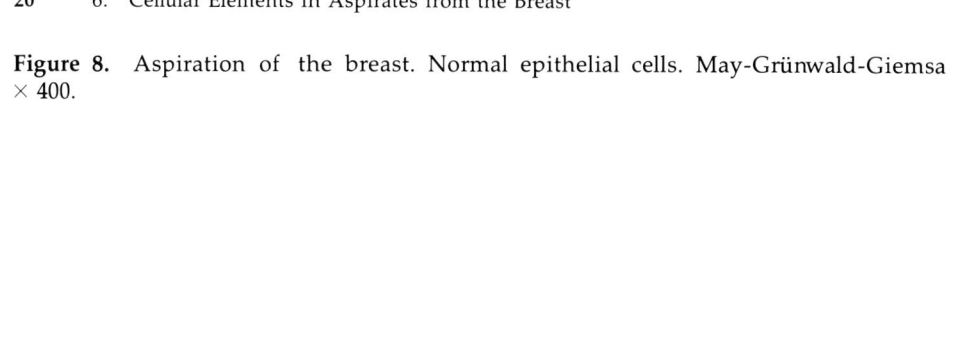

Figure 9.   Aspiration of the breast. Normal epithelial cells. Papanicolaou × 400.

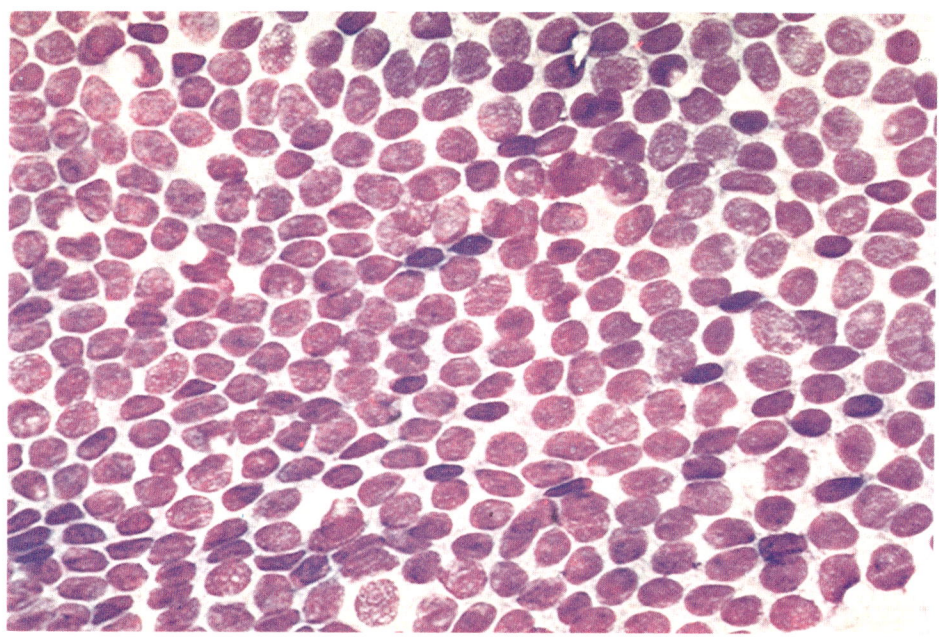

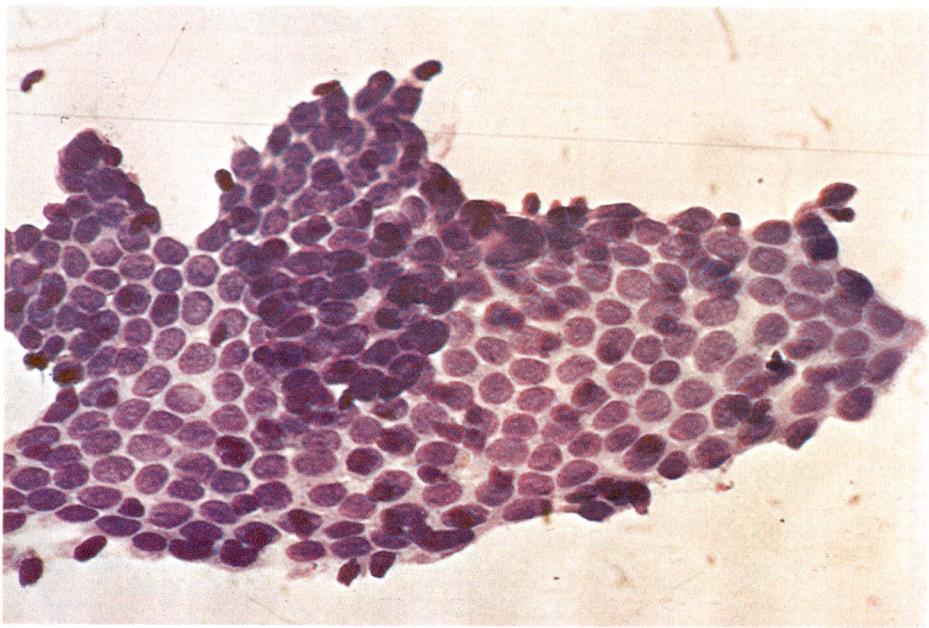

Figure 10.   Aspiration of the breast. Cluster of normal epithelial cells with uniformly round or oval nuclei. Papanicolaou × 1000.

Figure 11.   Cluster of normal epithelial cells from fibroadenoma. The nuclei are only slightly larger than erythrocytes. May-Grünwald-Giemsa × 400.

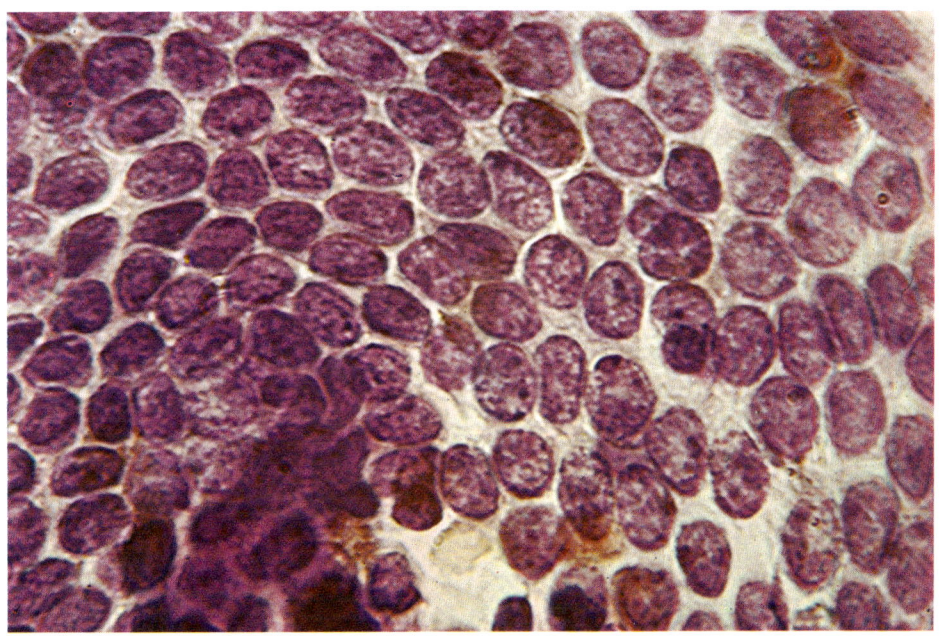

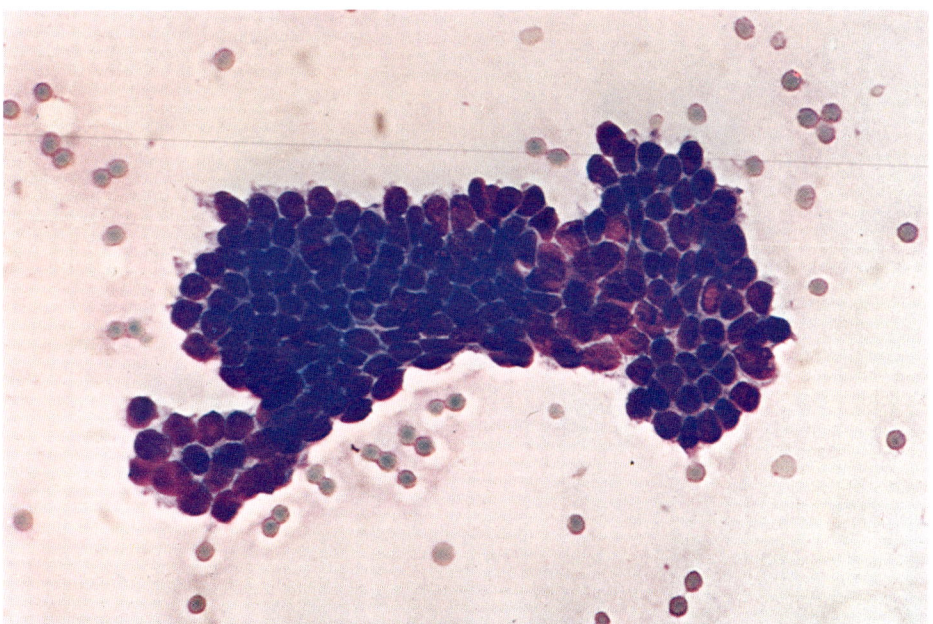

Figure 12. Epithelial cells with variation in size and shape. Superimposed are small, deeply staining naked bipolar nuclei, which are thought to represent myoepithelial cells. May-Grünwald-Giemsa × 400.

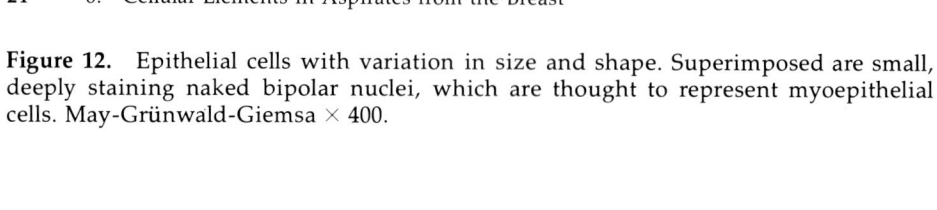

Figure 13. Three naked bipolar nuclei and a polymorphonuclear leukocyte (left). Cluster of epithelial cells (right). May-Grünwald-Giemsa × 1000.

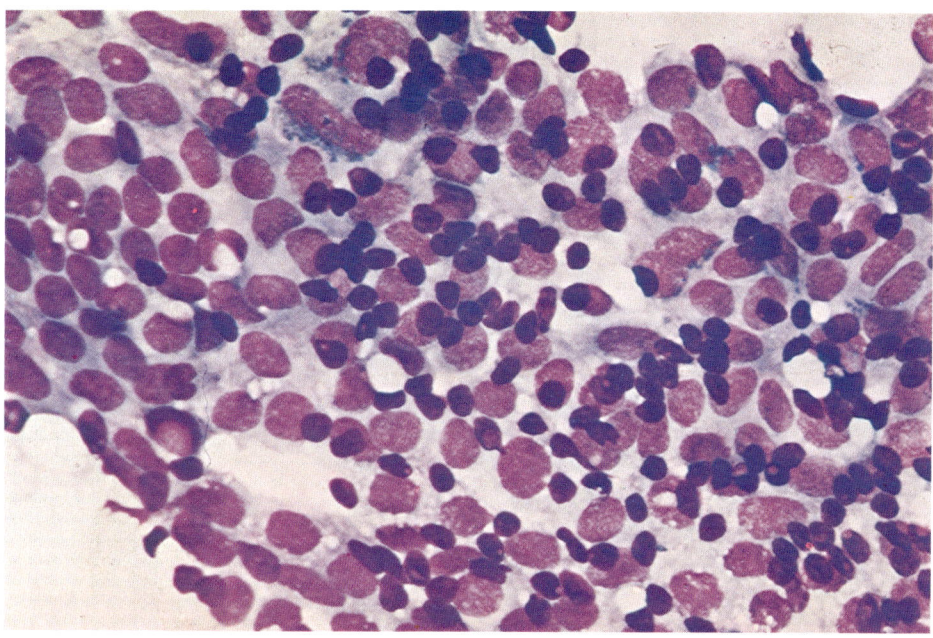

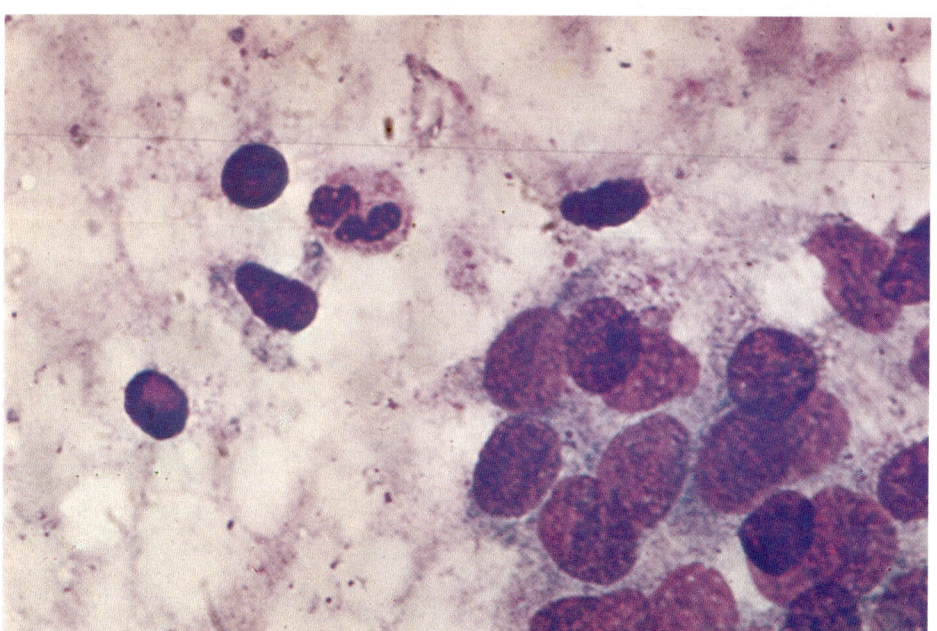

Figure 14.   Several naked bipolar nuclei, apocrine cell (right). May-Grünwald-Giemsa × 400.

Figure 15.   Group of apocrine cells with round, centrally located nuclei and prominent cell borders. May-Grünwald-Giemsa × 400.

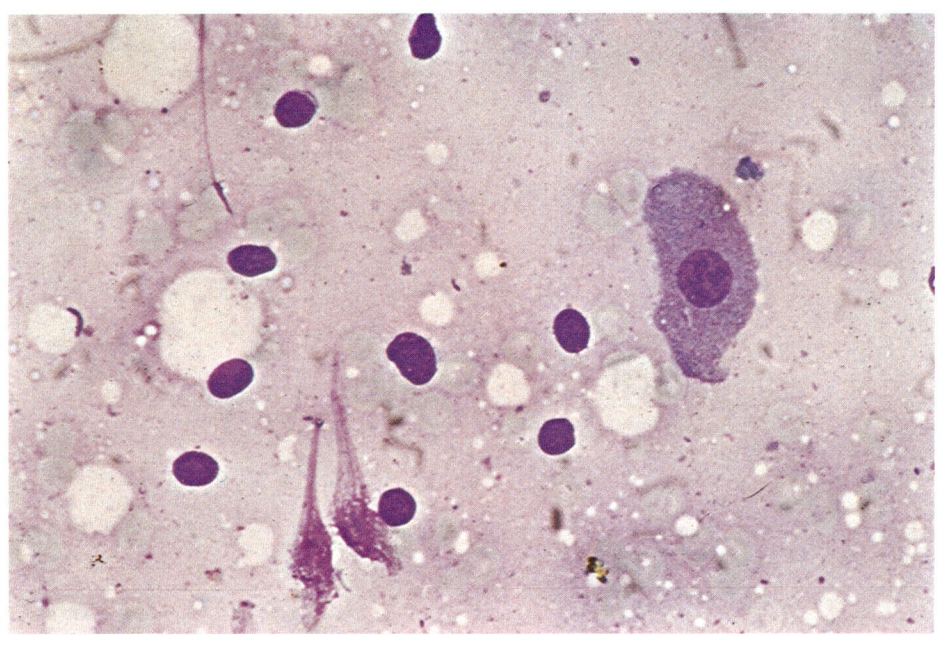

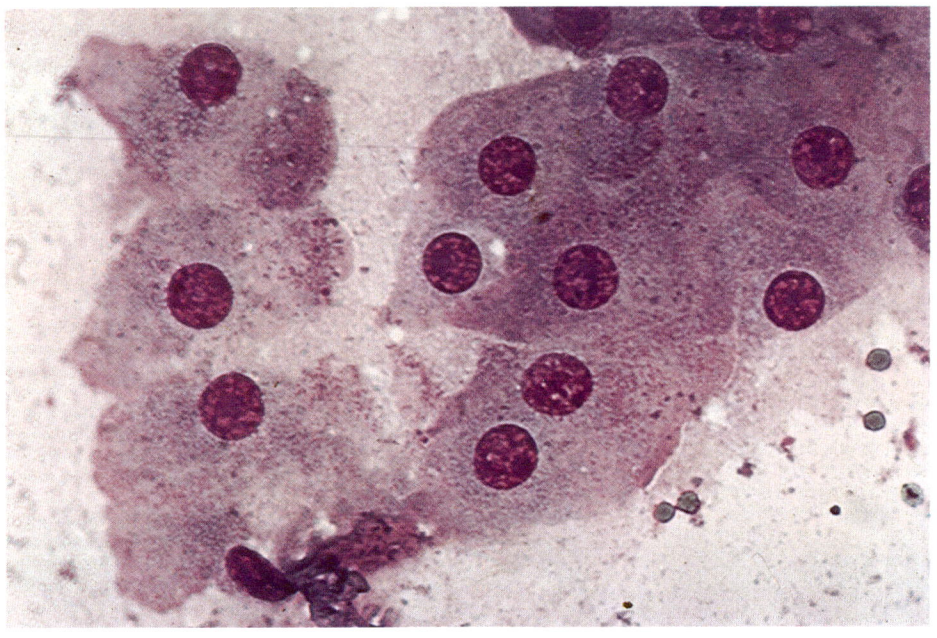

Figure 16.   Cluster of apocrine cells (bottom). Group of unremarkable epithelial cells (top right). May-Grünwald-Giemsa × 400.

Figure 17.   Apocrine cells with prominent granulation of the cytoplasm (center), surrounded by erythrocytes. May-Grünwald-Giemsa × 400.

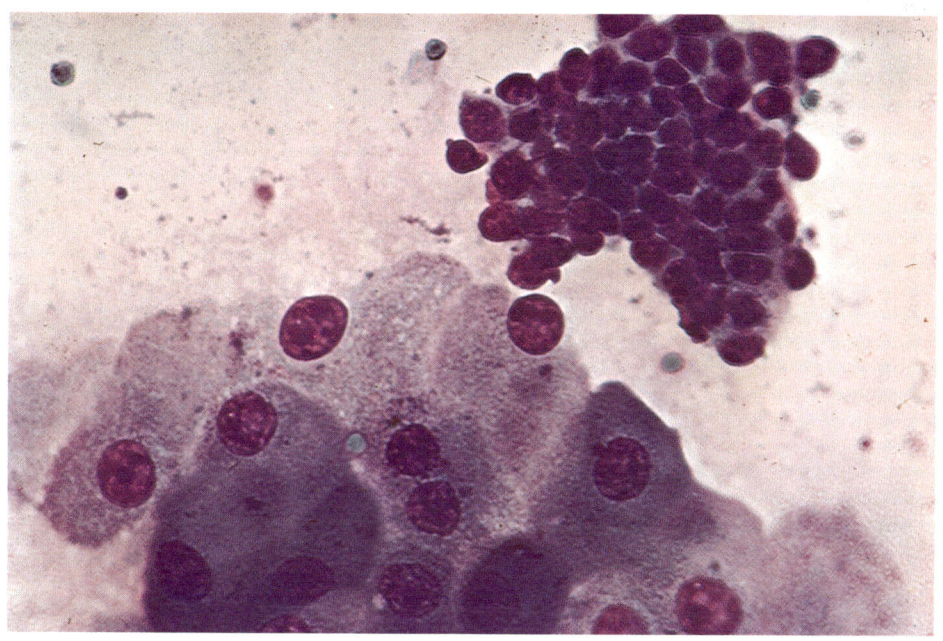

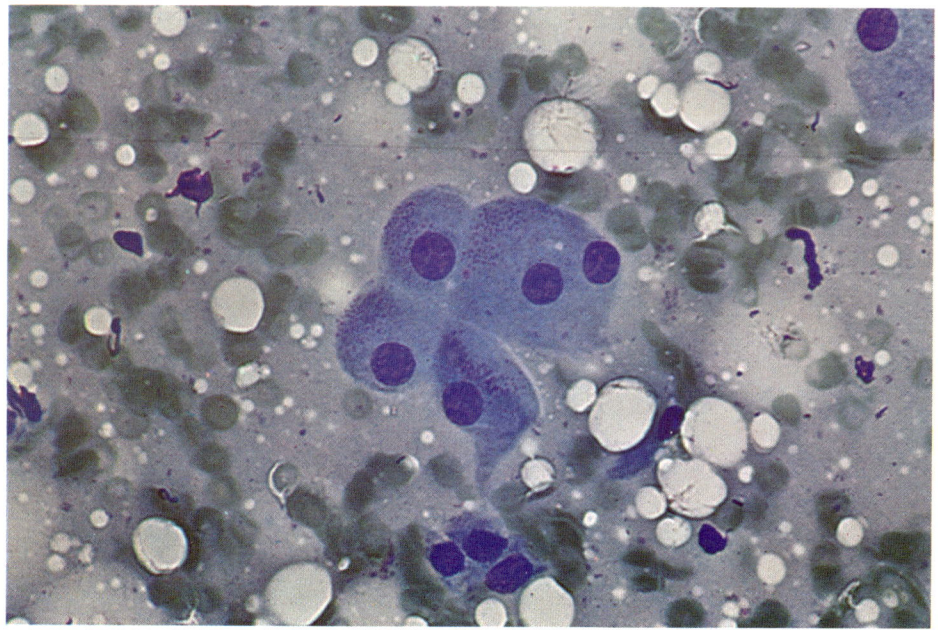

Figure 18. Large foam cells with eccentrically located nuclei. Several apocrine cells (top right). May-Grünwald-Giemsa × 400.

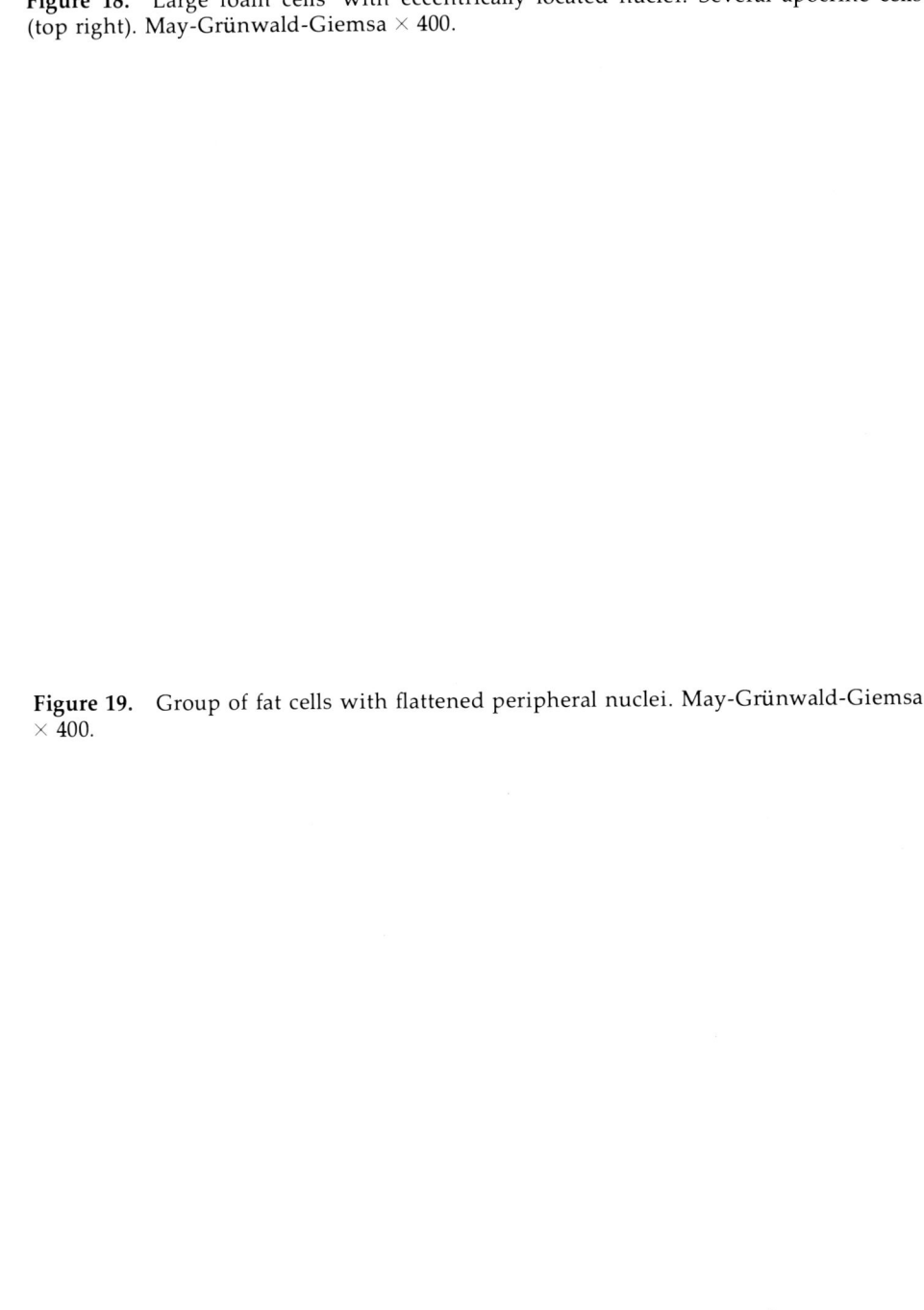

Figure 19. Group of fat cells with flattened peripheral nuclei. May-Grünwald-Giemsa × 400.

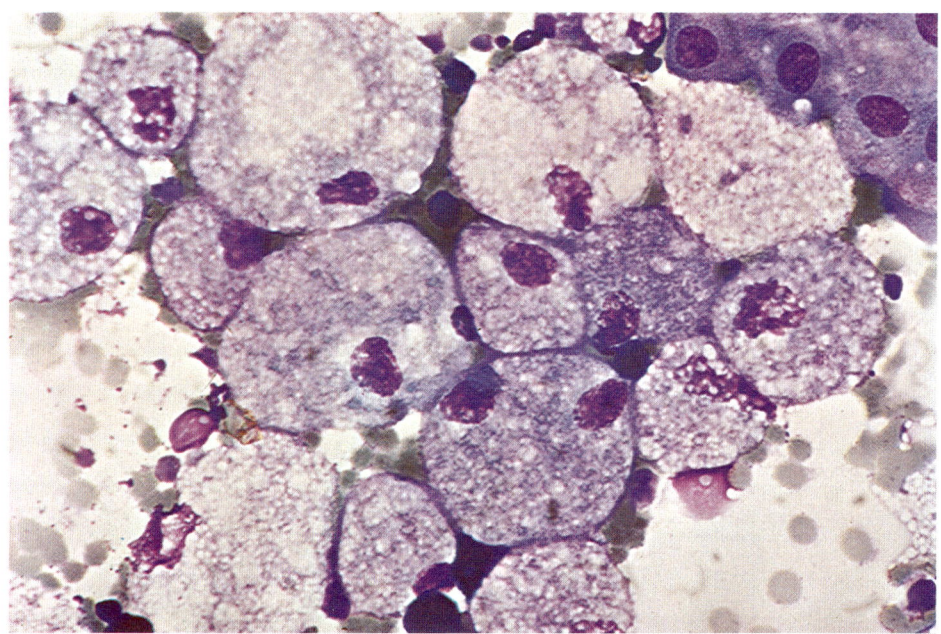

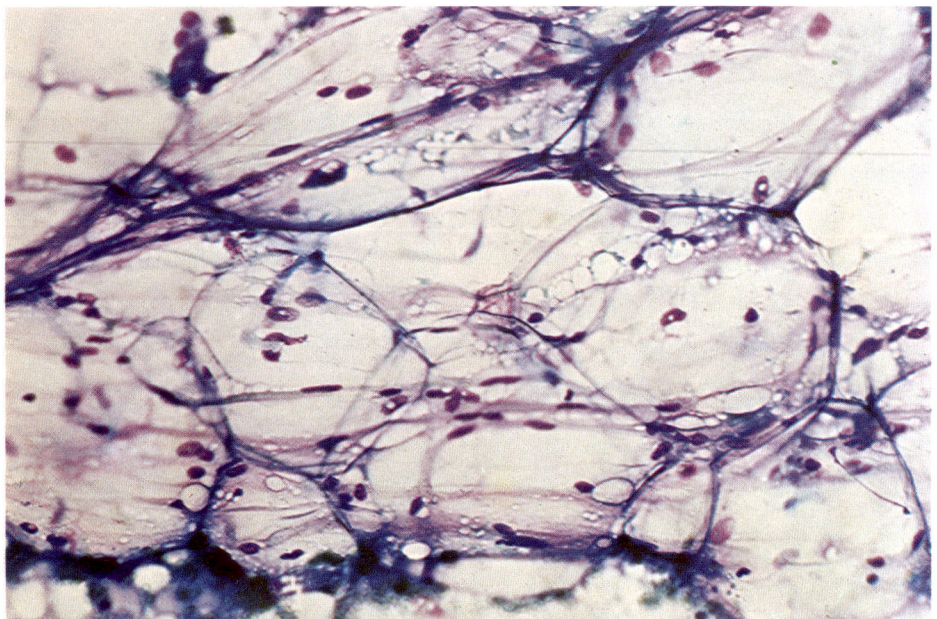

Figure 20. Large fragment of connective tissue containing oval and spindly-appearing fibrocytes. Apocrine cells with prominent granularity of the cytoplasm (bottom right). Large group of unremarkable epithelial cells (bottom left). Papanicolaou × 400.

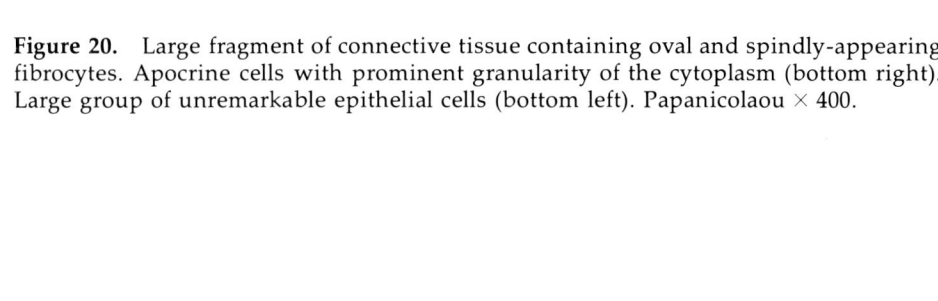

Figure 21. Aspirate from fibroadenoma containing fibrocytes, naked bipolar nuclei, and collagen. Small group of epithelial cells (bottom center). May-Grünwald-Giemsa × 400.

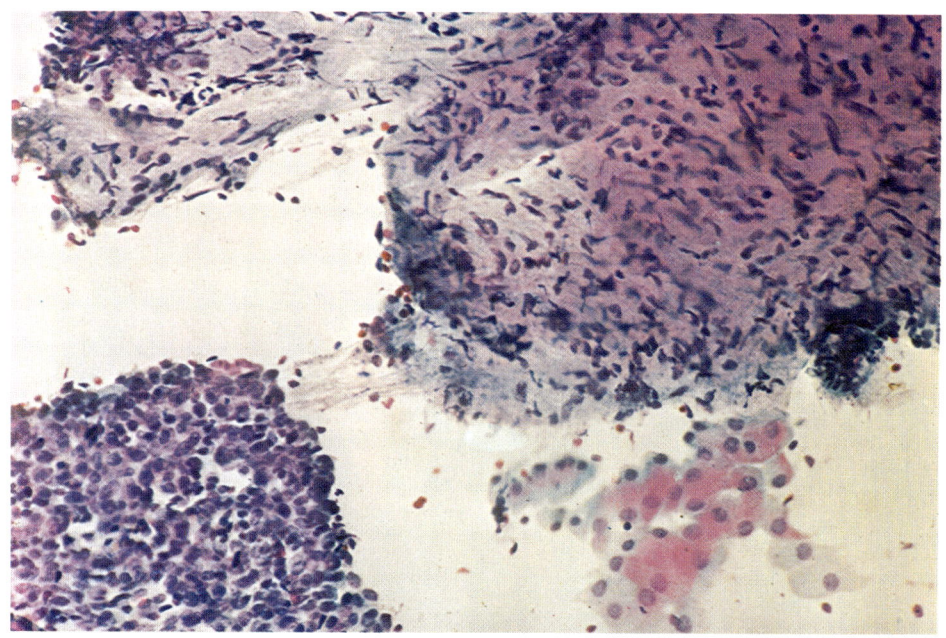

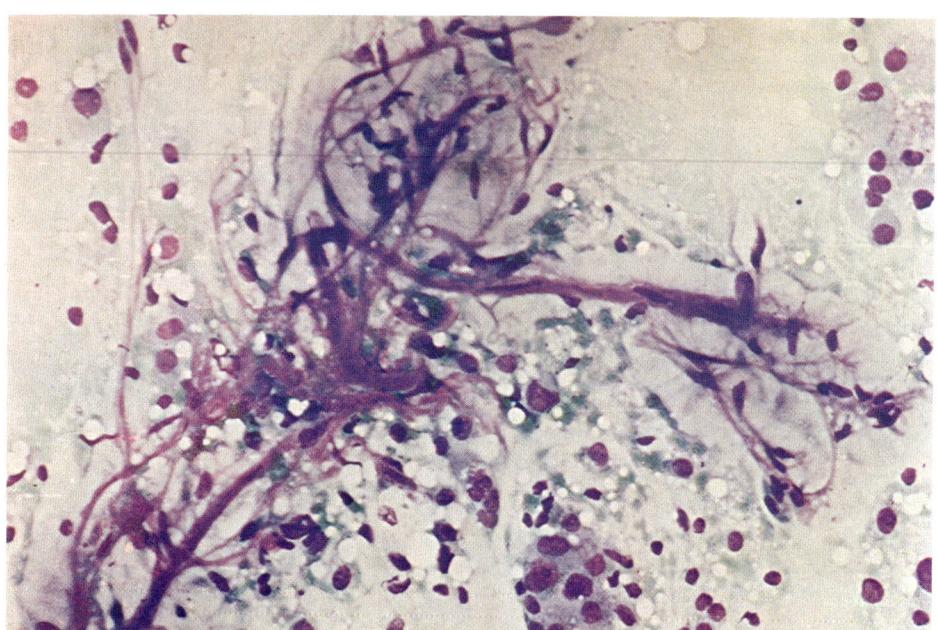

Figure 22.   Aspirate from mastitis. Giant cell surrounded by polymorphonuclear leuko-cytes. May-Grünwald-Giemsa × 400.

Figure 23.   Giant cell found in cyst fluid. Preparation by cytocentrifuge. May-Grün-wald-Giemsa × 400.

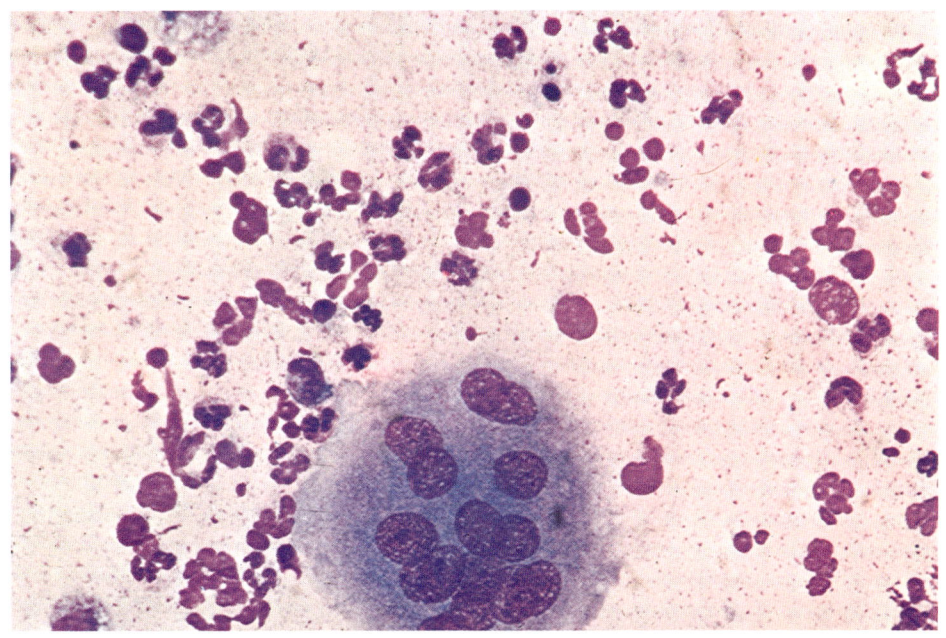

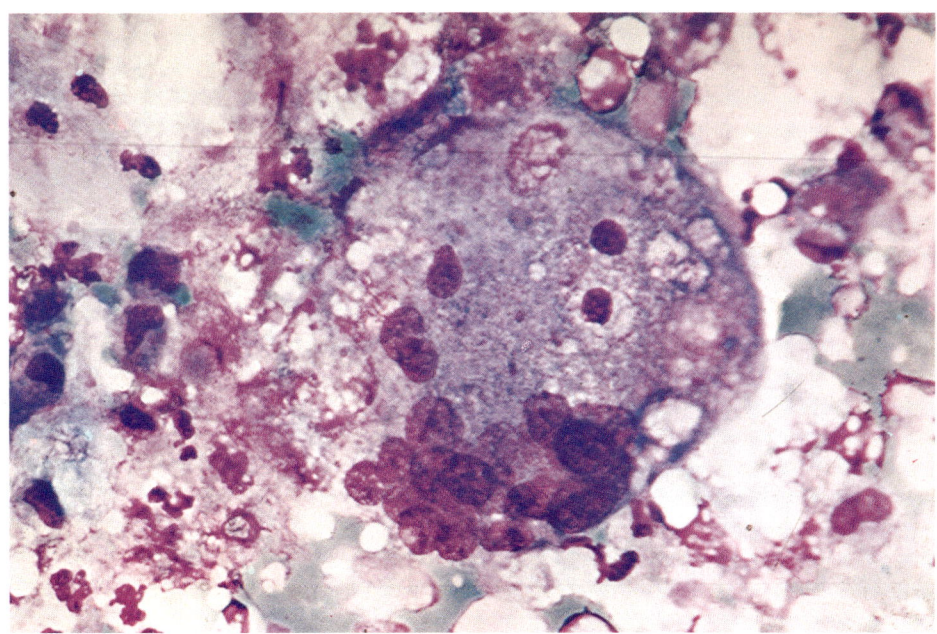

Figure 24.   Foreign body-type giant cell with loose, granular cytoplasm. May-Grünwald-Giemsa ×400.

Figure 25.   Giant cell in aspirate from a pregnant patient (32 weeks). May-Grünwald-Giemsa × 400.

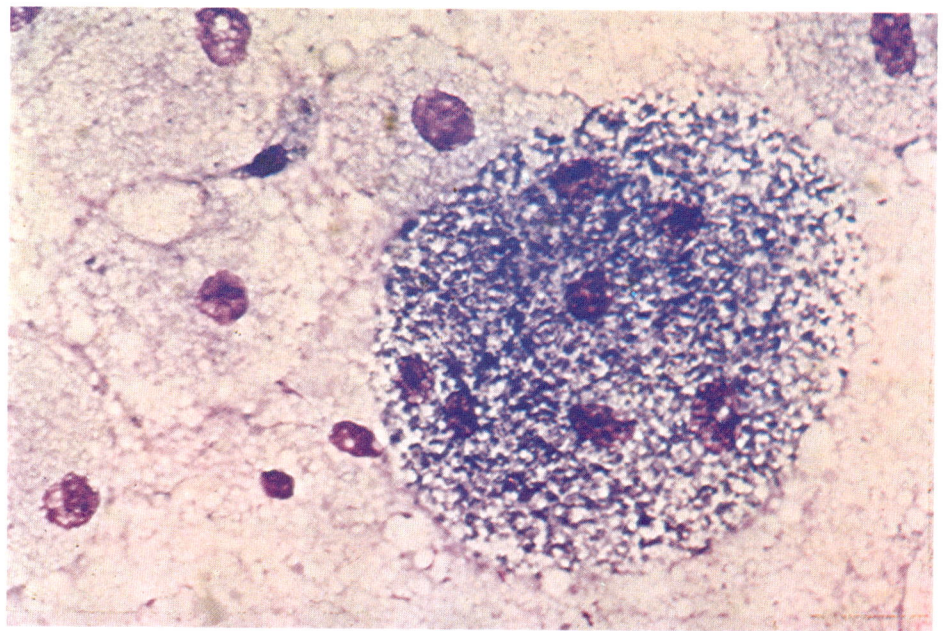

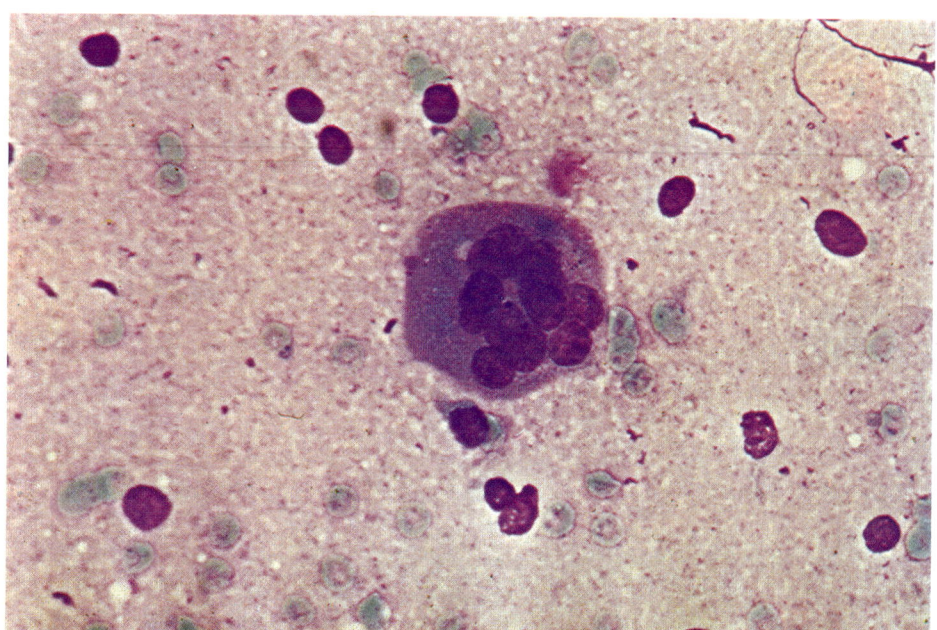

6.3.   Tumor cells

6.3.1.   Criteria of malignancy in fine needle aspirates

The cytological diagnosis of malignancy requires unequivocal evidence of the malignant nature of the cells. Therefore, at least two major criteria of malignancy should be present before such a diagnosis is considered. Some of the criteria that are well established in exfoliative cytology are less important in material obtained by needle aspiration.[22, 85, 98, 103] For example, alterations of the nuclear to cytoplasmic ratio cannot be evaluated, since cytoplasm is absent in most malignant cells. The cytological diagnosis of malignancy therefore relies almost exclusively on nuclear changes. In our material, a study of various criteria of malignancy was done in 120 cases of histologically confirmed carcinoma of the breast. The results indicated that (1) nuclear enlargement and pleomorphism are the most reliable criteria, and (2) in the majority of cases, these two criteria are sufficient to establish a definitive diagnosis of malignancy. The establishment of nuclear pleomorphism and enlargement as major criteria of malignancy in needle aspirates of the breast is generally accepted.[11, 17, 18, 20, 29, 35, 40, 42, 53, 61, 72, 83, 90, 98, 106, 112, 116]

6.3.2.   Nuclear size

Nuclei of tumor cells are several times larger than their benign counterparts. The nuclear diameter of malignant cells originating in the breast ranges from 12 to 40 μ. A rapid estimation of nuclear size can be performed easily by comparing the nucleus with erythrocytes measuring 7.5 μ in diameter (Fig. 26). Even in tumors with a rather uniform-appearing cell population, nuclear size usually allows correct identification. As an exception to this rule, the small cell variant of breast carcinoma usually contains nuclei of approximately the same size as benign epithelial cells (Fig. 28). Therefore, this cell type is easily misdiagnosed. Clues to the correct diagnosis are the presence of moderate pleomorphism and the loss of cellular cohesiveness.

6.3.3.   Pleomorphism

In the literature, the term "pleomorphism" is often used to indicate variation in nuclear shape as well as in nuclear size (anisokaryosis).[98] In carcinoma of the breast, both variations can be seen in the majority of malignant cells (Fig. 29). Occasionally, some degree of pleomorphism may be encountered in smears obtained from fibroadenoma or florid fibrocystic disease. However, numerous unequivocally bland nuclei generally are also found in specimens obtained from these benign conditions.

At low magnification, tumor cell nuclei appear surprisingly uniform in approximately 15 per cent of all malignant lesions of the breast. This picture is often found in well-differentiated carcinoma (Figs. 31 and 36). At higher magnification ($\times$ 400), irregularities of the nuclear membrane become quite obvious. These include thickening of the membrane, formation of clefts, and scalloping of the margins.

6.3.4. Nuclear to cytoplasmic ratio

Needle aspirates of the breast reveal the absence of cytoplasm in most malignant cells. Occasionally, a small rim of cytoplasm with hazy margins may be present (Fig. 35). It appears that the cytoplasm is destroyed during the process of aspiration. Thus, nuclear to cytoplasmic ratio cannot be applied as a criterion of malignancy in the majority of cells. Highly differentiated forms of carcinoma, however, may possess well-preserved cytoplasm (Fig. 37).

6.3.5. Nucleoli

Malignant cells often contain enlarged or multiple nucleoli (Fig. 32). However, the presence of prominent nucleoli should not be considered an indication of malignancy *per se*, since prominent nucleoli are regularly observed in benign proliferative conditions. Nucleoli of exceptionally large size are usually found only in malignancy.[98] On the other hand, absence of prominent nucleoli does not automatically exclude neoplastic disease. In May-Grünwald-Giemsa-stained material, nucleoli appear faintly bluish. In Papanicolaou-stained specimens, nucleoli appear darker and are easily visualized against a background of clear chromatin.

6.3.6. Atypical nuclear shape

As a result of impaired mitotic activity,[48] tumor cells may show multinucleation containing well-preserved nuclei as well as nuclear fragments of varying size (Fig. 38). In radiated tumor cells, extremely bizarre nuclear configurations may be observed, e.g., multilobulation or gross vacuolization (Fig. 39). Arrest of cellular division is thought to be the mechanism for the formation of tumor giant cells,[48] which are considerably larger than the usual tumor cells, reaching a diameter of 50 μ and larger.

In signet-ring cells, the nucleus is flattened and displaced into the periphery of the cell by a large cytoplasmic vacuole (Fig. 35). Signet-ring cells have been reported to occur more often in mucinous carcinoma, a finding that cannot be confirmed in our series.

6.3.7. Atypical mitoses

Atypical mitotic figures are only rarely observed. However, the number of normal-appearing mitoses is usually markedly increased in smears from malignant lesions (Fig. 35). (Because other criteria of malignancy, e.g., hyperchromasia and coarse granularity of the chromatin, are of less importance in needle aspirations of the breast, a detailed description is omitted.)

6.3.8 Indirect signs of malignancy

Loss of cellular cohesiveness and marked increase in cellularity are important indirect signs that indicate the presence of a malignant lesion. Loss of mutual cohesiveness is a prominent and well-known feature of cancer cells. Cell clusters dissociate as a result of loss of intercellular bridges, causing irregular distribution of the malignant nuclei with characteristic loss of polarity (Fig. 32).[83, 120] However, particularly in mucinous carcinoma, the mutual adherence of tumor cells may be preserved (Figs. 66 and 67). Occasionally, tumor cells form dense aggregates (Fig. 34). These clumps of cells should be examined very cautiously, since nuclear details are often obscured by superimposition.

Rich cellularity is another characteristic of smears obtained from malignant lesions of the breast.[72] The removal of tumor cells from the parent tissue is facilitated by the loss of mutual cohesiveness, especially in the presence of negative pressure during aspiration. As an exception to this rule, scirrhous carcinoma often yields only a few cells, making a definitive diagnosis impossible. In our series, poor cellularity was the reason for questionable or false negative cytological diagnosis in seven out of 68 scirrhous carcinomas. Histologically, the cells of scirrhous carcinoma are surrounded by dense fibrous stroma and therefore not easily released. Only multiple aspirations or the use of large bore needles may give better results.

The presence of erythrocytes or mucus in the aspirate has no special significance. Also, intranuclear vacuoles are more likely to represent nuclear degeneration than malignancy.

Figures 26 through 41 appear in the following section.

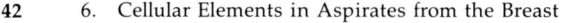

Figure 26. Large malignant nuclei in an aspirate from duct cell carcinoma. The large size can readily be appreciated in comparison with the size of erythrocytes. May-Grünwald-Giemsa × 400.

Figure 27. Malignant nuclei of intermediate size in an aspirate from duct cell carcinoma. May-Grünwald-Giemsa × 400.

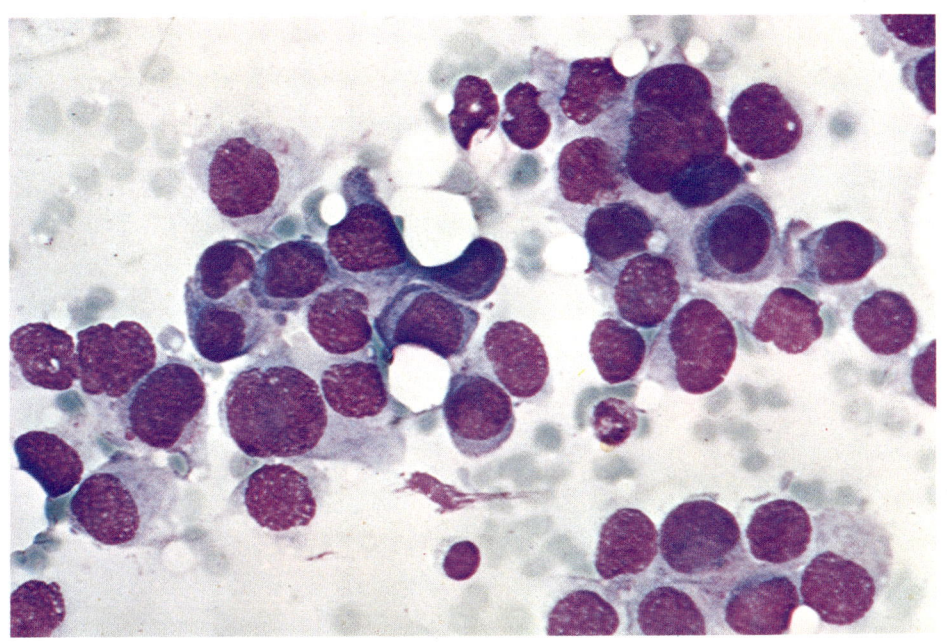

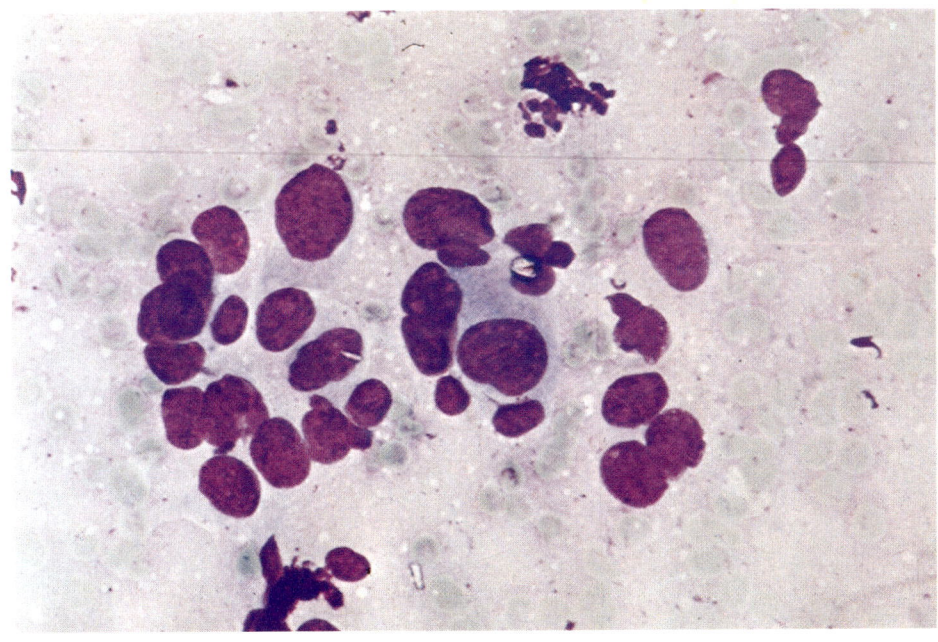

Figure 28.   Small cell variant of duct cell carcinoma. There is only a slight difference in size and shape between these nuclei and their benign counterparts. This rare cell type therefore is easily misdiagnosed. May-Grünwald-Giemsa × 400.

Figure 29.   Prominent nuclear pleomorphism in an aspirate from intraductal carcinoma. Papanicolaou × 400.

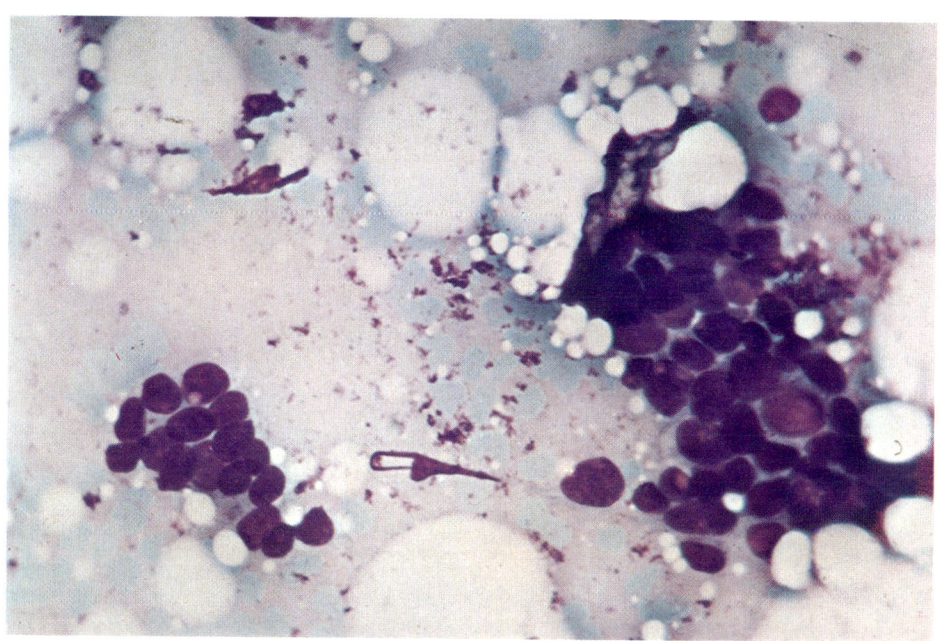

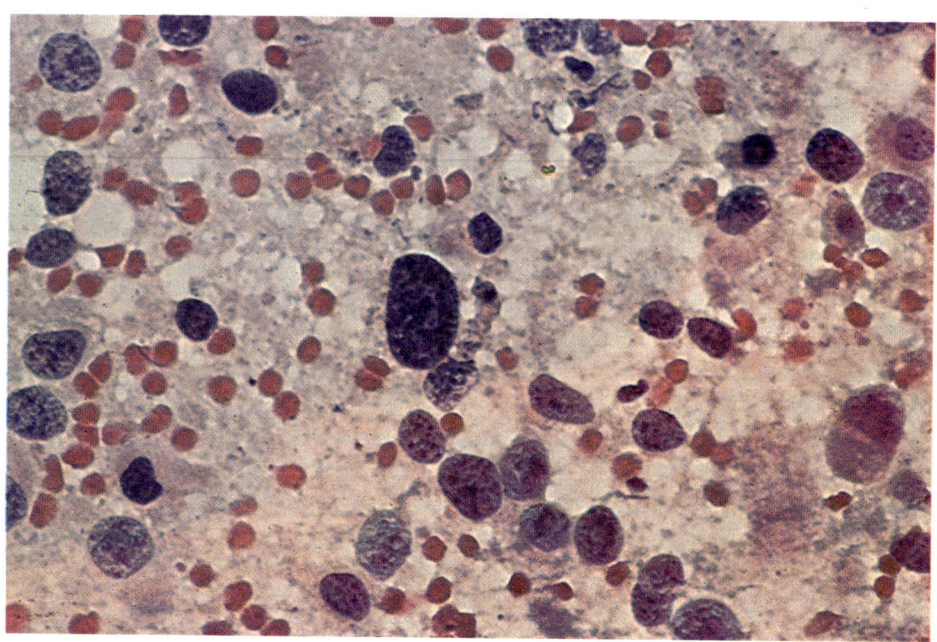

Figure 30.   Marked pleomorphism and increase in nuclear size in an aspirate from duct cell carcinoma. May-Grünwald-Giemsa × 400.

Figure 31.   Relatively uniform-appearing tumor cells of medullary carcinoma. May-Grünwald-Giemsa × 400.

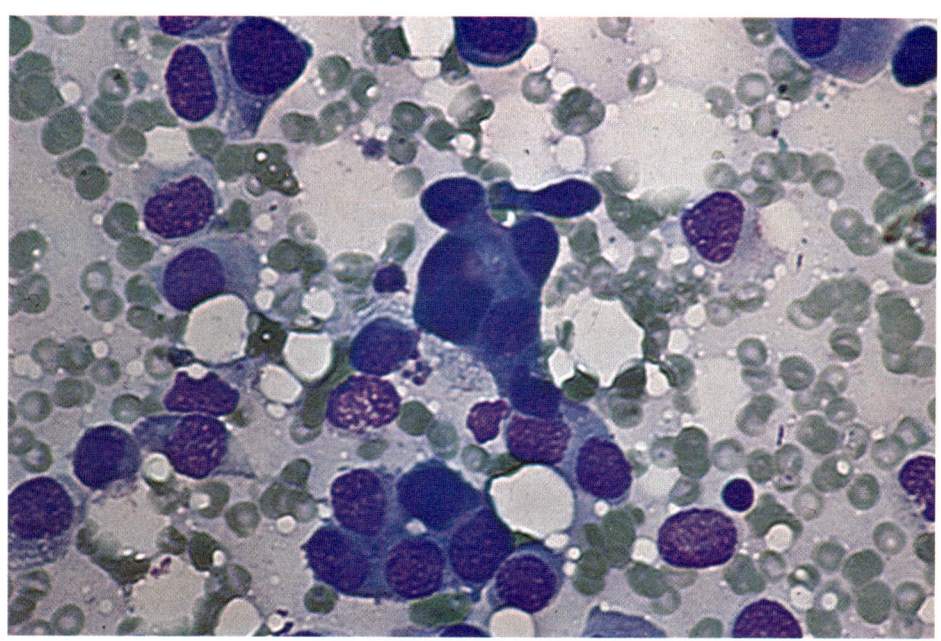

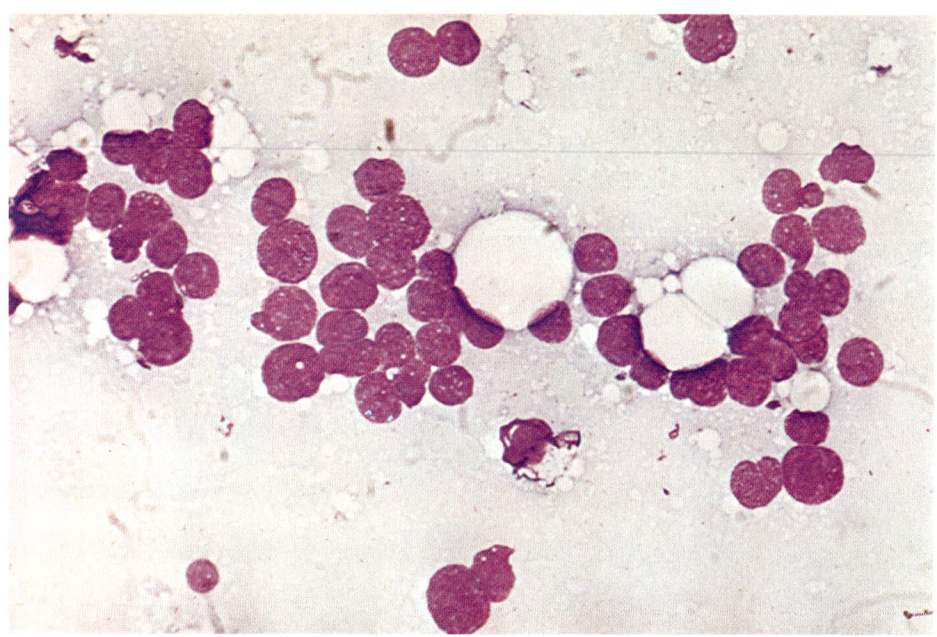

Figure 32. Complete dissociation of nuclei and prominent enlargement of nucleoli in duct cell carcinoma. May-Grünwald-Giemsa × 400.

Figure 33. Cluster of tumor cells with prominent pleomorphism in duct cell carcinoma. May-Grünwald-Giemsa × 400.

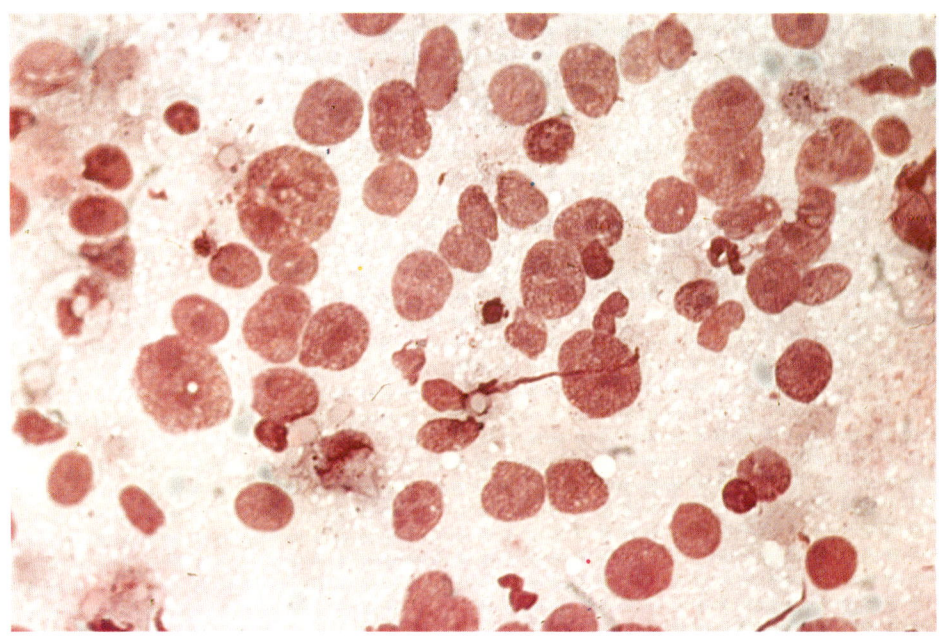

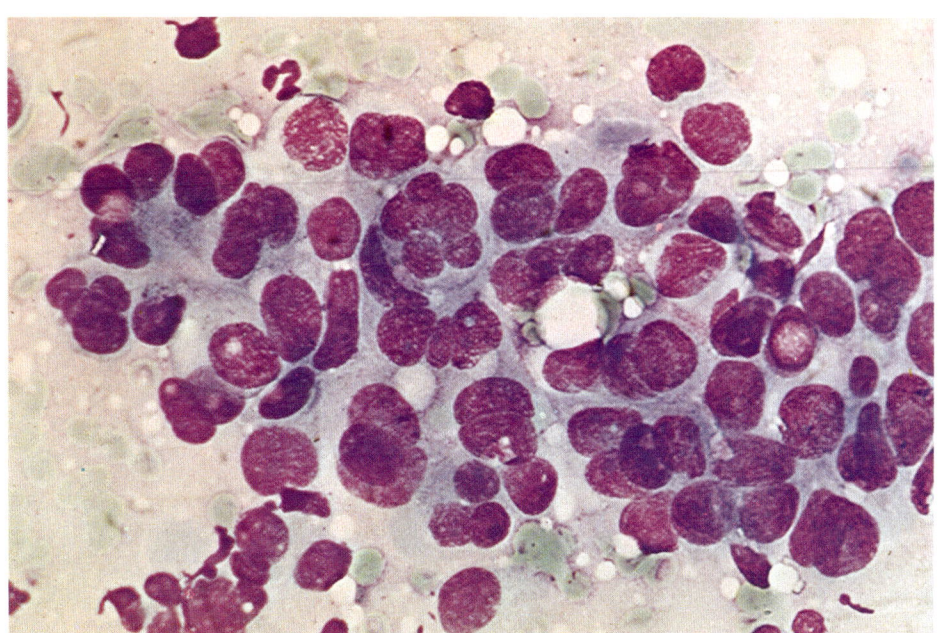

Figure 34.   Nuclear crowding. The malignant character of the cells is more readily appreciated at the margins of the cluster. Duct cell carcinoma. May-Grünwald-Giemsa × 400.

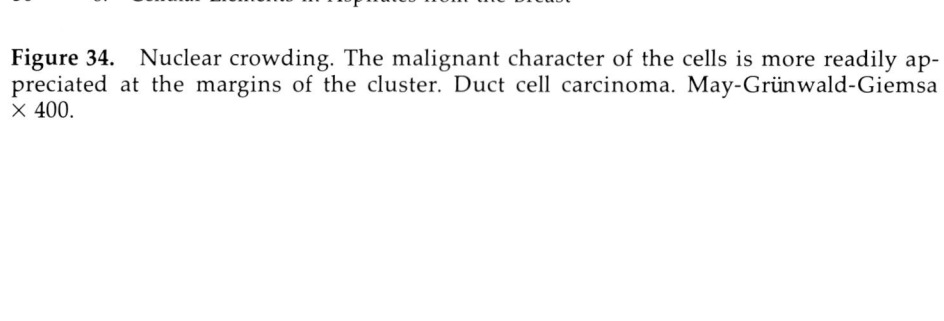

Figure 35.   Pleomorphism, signet-ring cell formation, mitosis, and alteration of the nuclear-cytoplasmic ratio. Rapidly growing medullary carcinoma. Papanicolaou × 400.

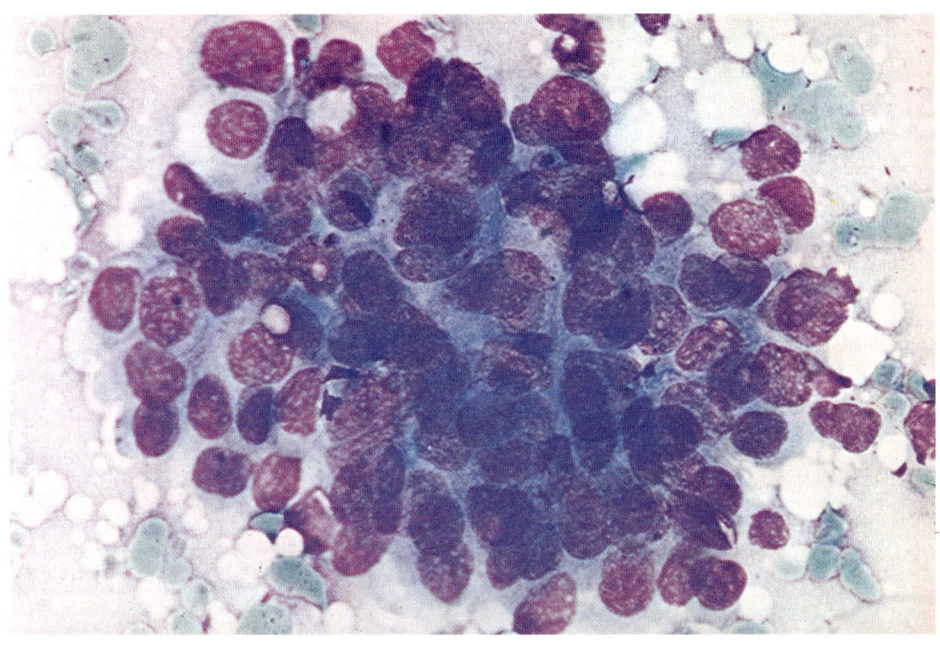

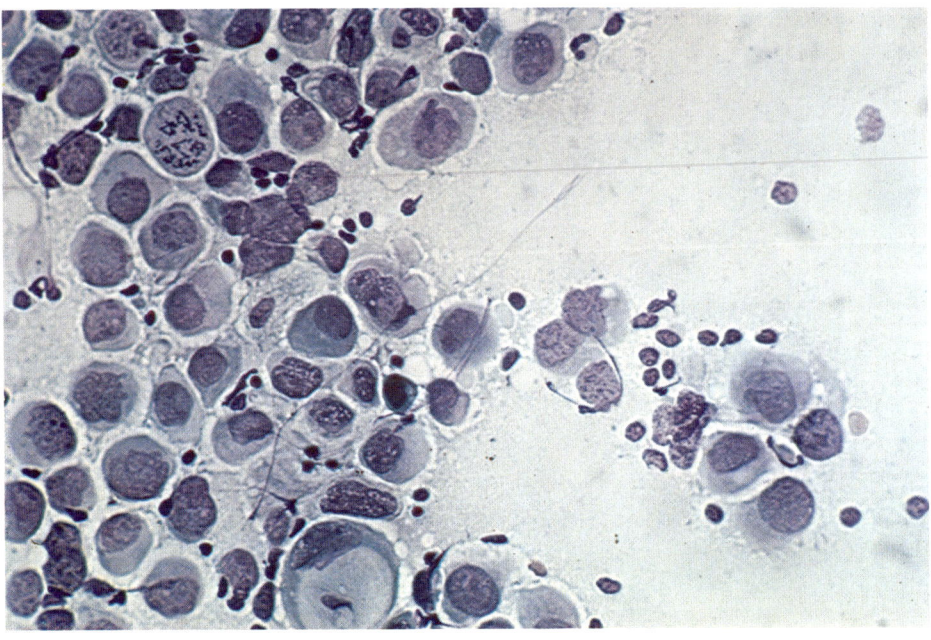

Figure 36. Well-differentiated duct cell carcinoma with uniform-appearing nuclei. May-Grünwald-Giemsa × 400.

Figure 37. Cells of apocrine carcinoma. Papanicolaou × 400.

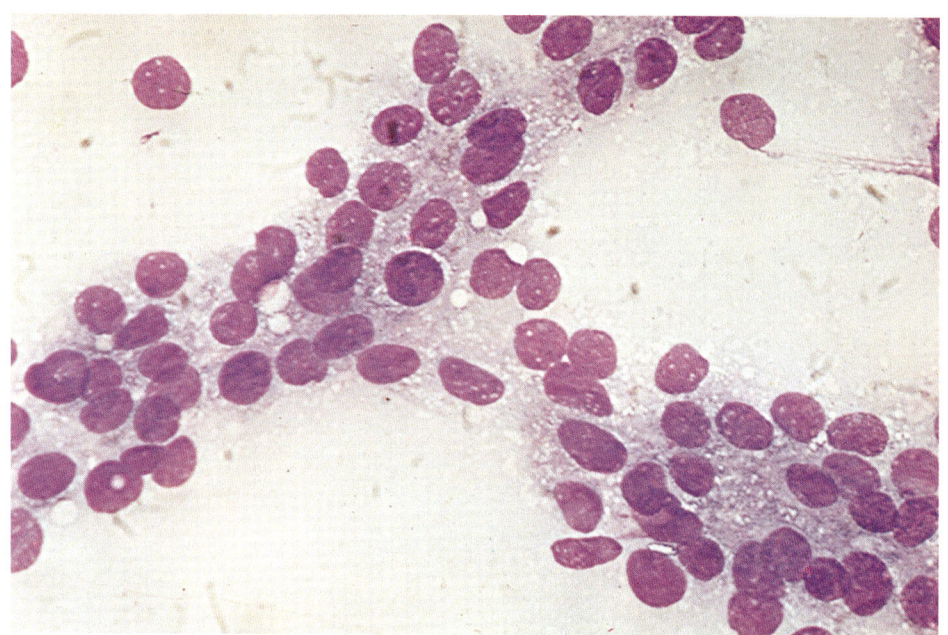

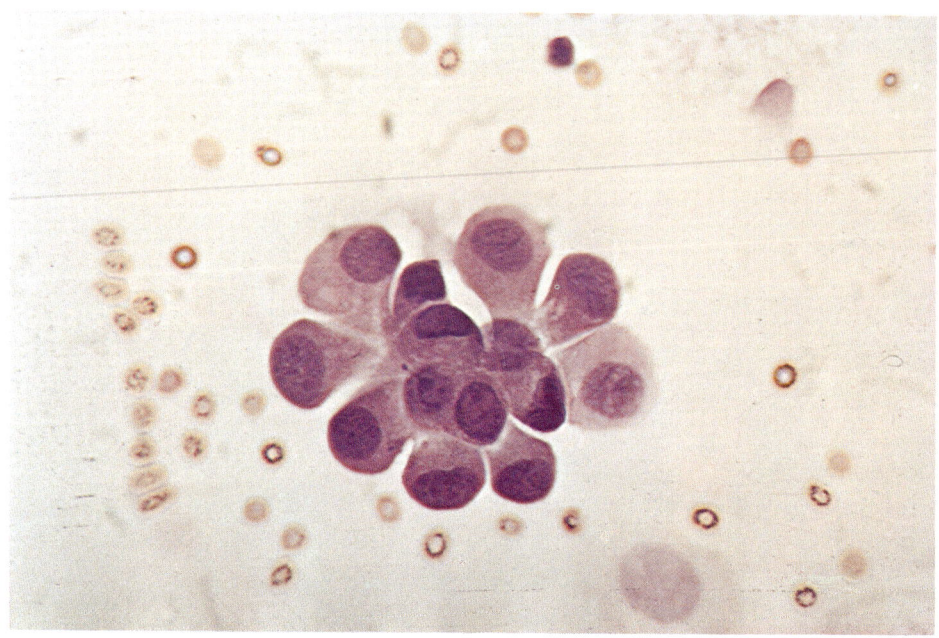

Figure 38. Tumor giant cells in an aspiration from a metastatic axillary lymph node. Medullary carcinoma. Papanicolaou × 400.

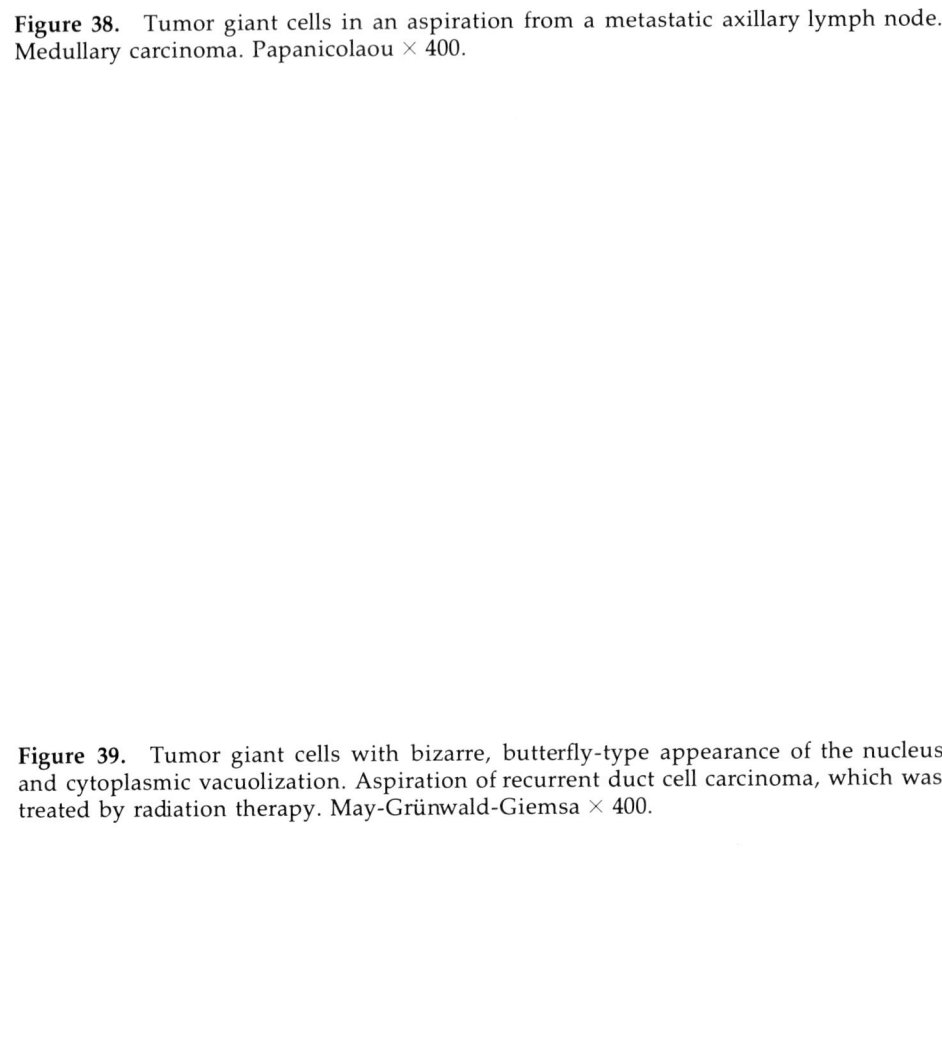

Figure 39. Tumor giant cells with bizarre, butterfly-type appearance of the nucleus and cytoplasmic vacuolization. Aspiration of recurrent duct cell carcinoma, which was treated by radiation therapy. May-Grünwald-Giemsa × 400.

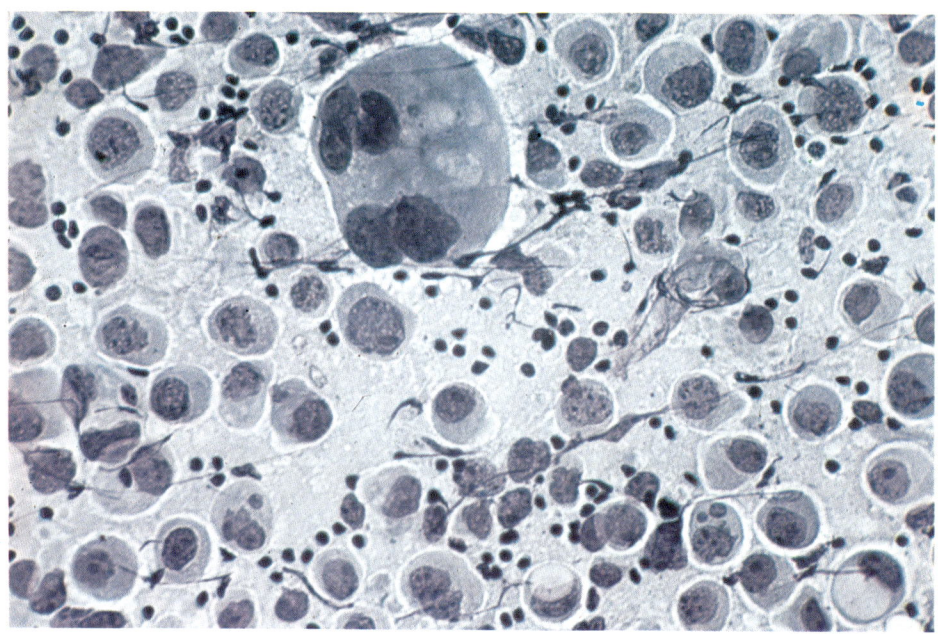

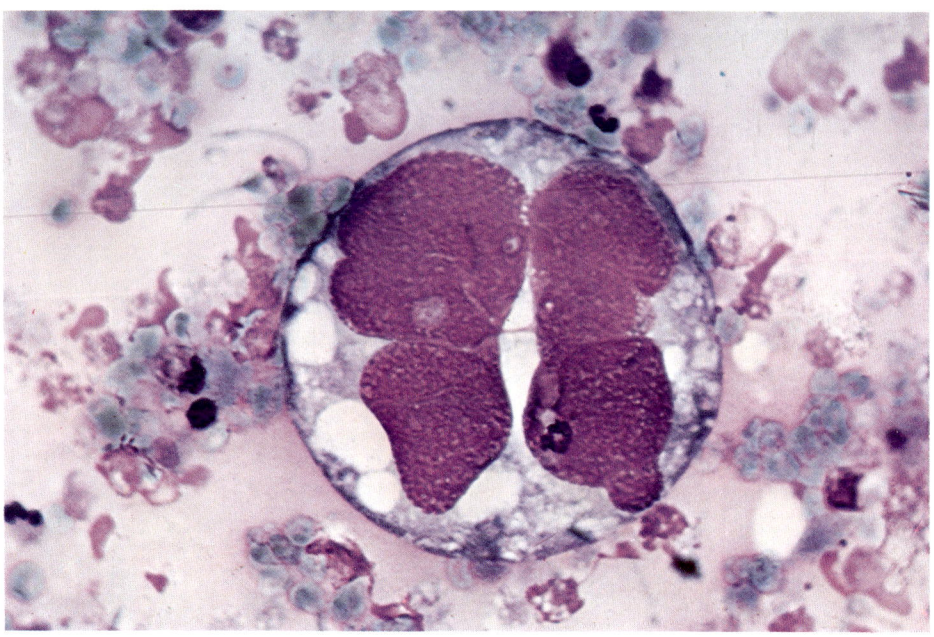

Figure 40.   Cluster of malignant cells with vacuolar degeneration of the cytoplasm. Nuclear debris and erythrocytes in the background. Aspiration of recurrent duct cell carcinoma, which was treated by radiation therapy. May-Grünwald-Giemsa × 400.

Figure 41.   Malignant cells in aspirated cyst fluid with marked pleomorphism and pigment granules. May-Grünwald-Giemsa × 400.

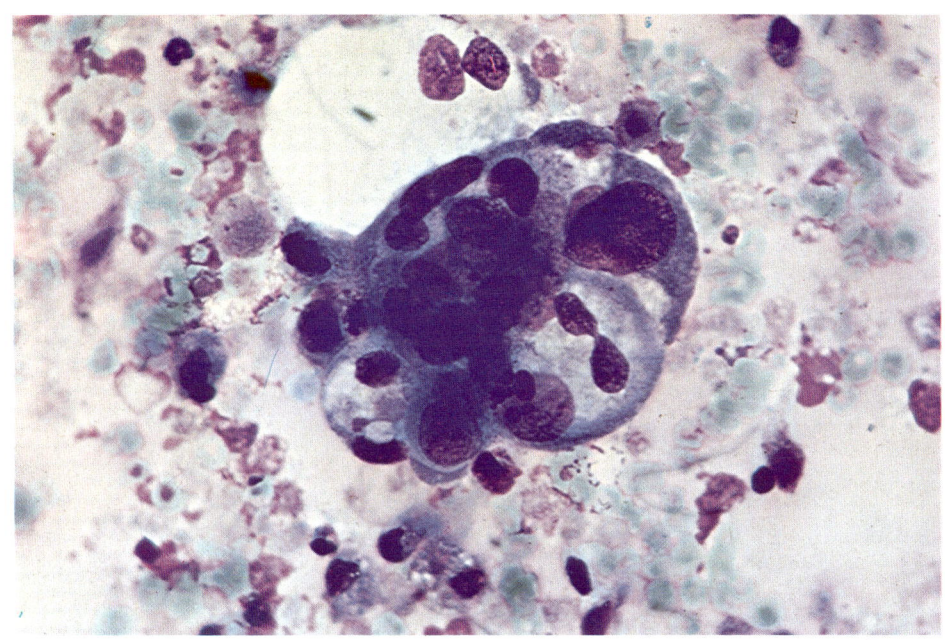

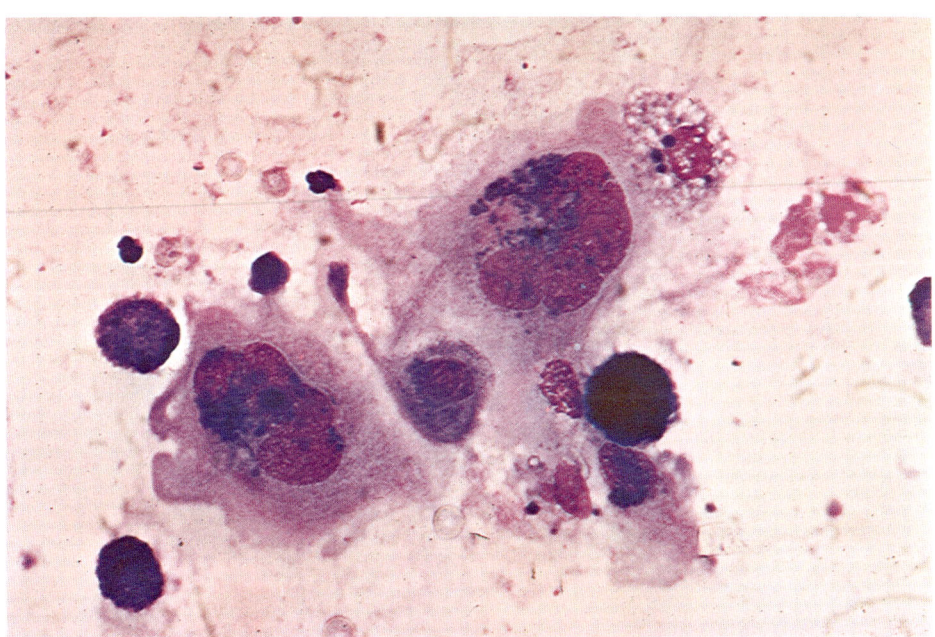

7. Evaluation of the smear

7.1. Classification

We classify smears obtained by fine needle aspiration of the breast as follows: (1) unremarkable smear, (2) suspicious smear, (3) tumor cells present, and (4) unsatisfactory specimen.

Papanicolaou's classification[85] is used by several authors with various modifications.[8, 10, 13, 27] Some of the descriptive terms used, such as "atypical cells," "hyperplasia," "slightly abnormal cells," "malignant appearance," and "suspicion of tumor," correspond to our category of "suspicious smear."

Unremarkable smear

Generally, smears classified as unremarkable should not show any substantial cellular enlargement or nuclear pleomorphism. However, a mild degree of atypia may occasionally be present. The appearance of the cells usually conforms to the criteria outlined in the section on normal cellular elements of the breast.

Some of the features still considered to be within normal limits are:

1. Marked cellularity, e.g., in aspirates from fibroadenoma. Smears from fibroadenoma may also show considerable variation in nuclear size. However, the nuclear diameter of benign duct cells rarely exceeds 11 μ (in comparison, the diameter of erythrocytes is 7.5 μ). Similarly, nuclei of apocrine metaplasia may show considerable variation in nuclear size.

2. Vacuolization of the cytoplasm, which occasionally may be observed in otherwise unremarkable duct and myoepithelial cells.

3. Marked nuclear vacuolization, multinucleation of foam cells, and enlargement of duct cells with prominent nucleoli. All these changes may be induced by hormonal stimulation, as occurs in pregnancy.

Suspicious smear

Smears classified as suspicious do not permit a definitive diagnosis on the basis of cytological findings alone. In this situation, excisional biopsy is necessary. Occasionally, special circumstances, such as pregnancy or inflammation, may explain the abnormal presentation. Cytologically, one of the following pictures may be found:

1. Isolated mild enlargement and pleomorphism of nuclei may be present in poorly cellular specimens.

2. Uniformly enlarged nuclei with prominent nucleoli may be found in aspirates from inflammatory or foreign body reactions or after hormonal therapy.

3. Occasional marked nuclear enlargement and moderate pleomorphism

may occur in otherwise unremarkable specimens. Fibroadenoma, florid fibro-cystic disease, or lobular carcinoma *in situ* may present in this way.

4. Because of their large size and prominent nuclear pleomorphism, histio-cytes in large numbers may be confused with malignant cells. However, the presence of a rim of pale, finely vacuolated cytoplasm with ill-defined borders should enable the cytologist to identify the cells accurately.

5. The small cell variant form of breast carcinoma is extremely difficult to diagnose because the nuclei do not differ substantially in size from those of benign epithelial cells. Fortunately, this cell type is uncommon. In our series, only two examples were found among 333 carcinomas of the breast.

6. Finally, approximately 15 per cent of all carcinomas of the breast show a surprisingly monomorphous picture in needle aspirates, often causing difficul-ties in cytological interpretation. In this context, acinic cell carcinoma should be mentioned.[122] Well-differentiated forms of this cell type may appear deceptive-ly similar to aspirates obtained during pregnancy and lactation. This cell type may be found in aspirations from duct cell carcinoma and medullary carcinoma.

These are some of the conditions frequently encountered in fine needle as-piration of the breast that may cause problems or even erroneous diagnoses. In these instances, surgical biopsy is recommended for rapid and reliable clarifica-tion.

In our material, fibroadenoma was found histologically in 35 per cent of all cases that were considered to be inconclusive cytologically. The tendency to epithelial proliferation and the common occurrence of fibroadenoma are proba-ble reasons for this lesion to predominate in this category. Fibrocystic disease was found in 30 per cent of cases as an underlying condition. Fat necrosis, gran-ulation tissue, and lobular carcinoma *in situ* were only occasional causes for a cytologically suspicious presentation.

Tumor cells present

The diagnosis "tumor cells present" represents an unequivocal statement of the malignant nature of the underlying lesion, indicating that it is either a pri-mary carcinoma or sarcoma, or a metastatic or recurrent malignancy. Usually, malignant cells show marked deviation from the normal appearance, often en-abling the examiner to make the diagnosis of malignancy at first glance. To con-firm this impression, the aspirates of 121 histologically proved cases of carcino-ma of the breast were reexamined. In 85 per cent the diagnosis of malignancy was immediately obvious, making further study unnecessary. Careful examina-tion of the entire slide was necessary in approximately 11 per cent before it could be determined that the lesion was malignant. In approximately 4 per cent, even a close scrutiny of the entire cell population did not provide enough evi-dence for an unequivocal diagnosis of malignancy. Evidence from clinical ex-amination, mammography, and thermography, as well as comparison with slides from the files, were necessary to support a diagnosis of malignancy.

It should be acknowledged that needle aspiration cytology in this situation has its limitations. It seems prudent to consider such specimens as cytologically inconclusive.

Unsatisfactory specimen

Reasons for rejection of a specimen as unsatisfactory are: poor preparation of the smear, poor fixation, poor staining technique, clumping of cells, insufficient identification of the specimen, and non-representative number of cells, as in aspirations from solid tumors.

Smears containing few or no cells should not be dismissed as "unremarkable." Poor cellularity may often be explained by the virtual absence of fluid in a small cyst that was aspirated or by marked fibrosis of the lesion. However, it is generally advisable to repeat the aspiration with a large bore needle or to perform a biopsy on the lesion if this is clinically indicated.

7.2. Report and documentation

In Scandinavia, the aspiration of the lesion as well as the preparation and interpretation of the smear usually is performed by the cytopathologist himself. This is considered an ideal situation.[10, 34, 98, 113] In Germany and many other countries, this arrangement is not always possible. Hence, the farther removed the aspiration of the lesion from the cytological interpretation of the smear, the better the communication between aspirating clinician and cytopathologist must be.[30] Clinicians performing fine needle aspirations should receive formal training in all technical aspects of the procedure. At the same time, detailed information should be provided for the cytopathologist who handles the material in the laboratory. In our department, a request sheet is used (Fig. 42) that provides information on the most important points, i.e., pertinent data from the history (primary, recurrent, or metastatic disease, chemotherapy, radiation therapy, pregnancy), results of the clinical examination (localization of the lesion and palpatory findings), and results of additional examinations (mammography and thermography).

The cytological interpretation is documented on Part 2 of the request sheet (Fig. 43), which is kept in our files. Part 3 is sent back to the requesting physician and contains the final diagnosis (Fig. 44). For legal and statistical purposes, a detailed description of the material present in the smear should be recorded. In our department, all cellular and non-cellular elements present in the aspirate are checked off on a schematic list in Part 2 of the request sheet (Fig. 43). Additional space is provided at the bottom for remarks and unusual findings. Cellular elements listed include epithelial cells of duct or lobular origin, myoepithelial cells, apocrine cells, foam cells, fat cells, fibrocytes, giant cells, and tumor cells.

For duct cells and particularly for tumor cells, additional information is recorded. Benign duct cells usually contain a round or oval nucleus and a scant rim of cytoplasm. The cell size rarely exceeds 8 μ in diameter. However, considerable deviation from this typical appearance may occur.

In order to investigate atypical presentations of duct cells and correlate these with histological findings, three grades of cellular atypia have been tentatively formed and are now under investigation:

Grade I: Moderate pleomorphism of duct cells. Numerous myoepithelial cells present.

Grade II: Pleomorphism of duct cells. Marked nuclear enlargement observed only occasionally. Myoepithelial cells present.

Grade III: Pleomorphism of duct cells. Marked nuclear enlargement observed frequently. Myoepithelial cells absent or sparse.

If tumor cells are present, their approximate number is estimated. In addition, information is recorded about the arrangement of cells, nuclear size, degree of pleomorphism, and amount of cytoplasm. The presence of special features, e.g., signet-ring cells, multinucleated cells, or mitotic figures, is noted. Background material, such as mucus, necrotic debris, erythrocytes, and leukocytes, is also noted.

Figures 42, 43, and 44 follow.

Hospital of the
Johann Wolfgang Goethe University
Department of Obstetrics and Gynecology
Division of Clinical Cytology

Number of slides: _____ Date: _____ Log No: _____

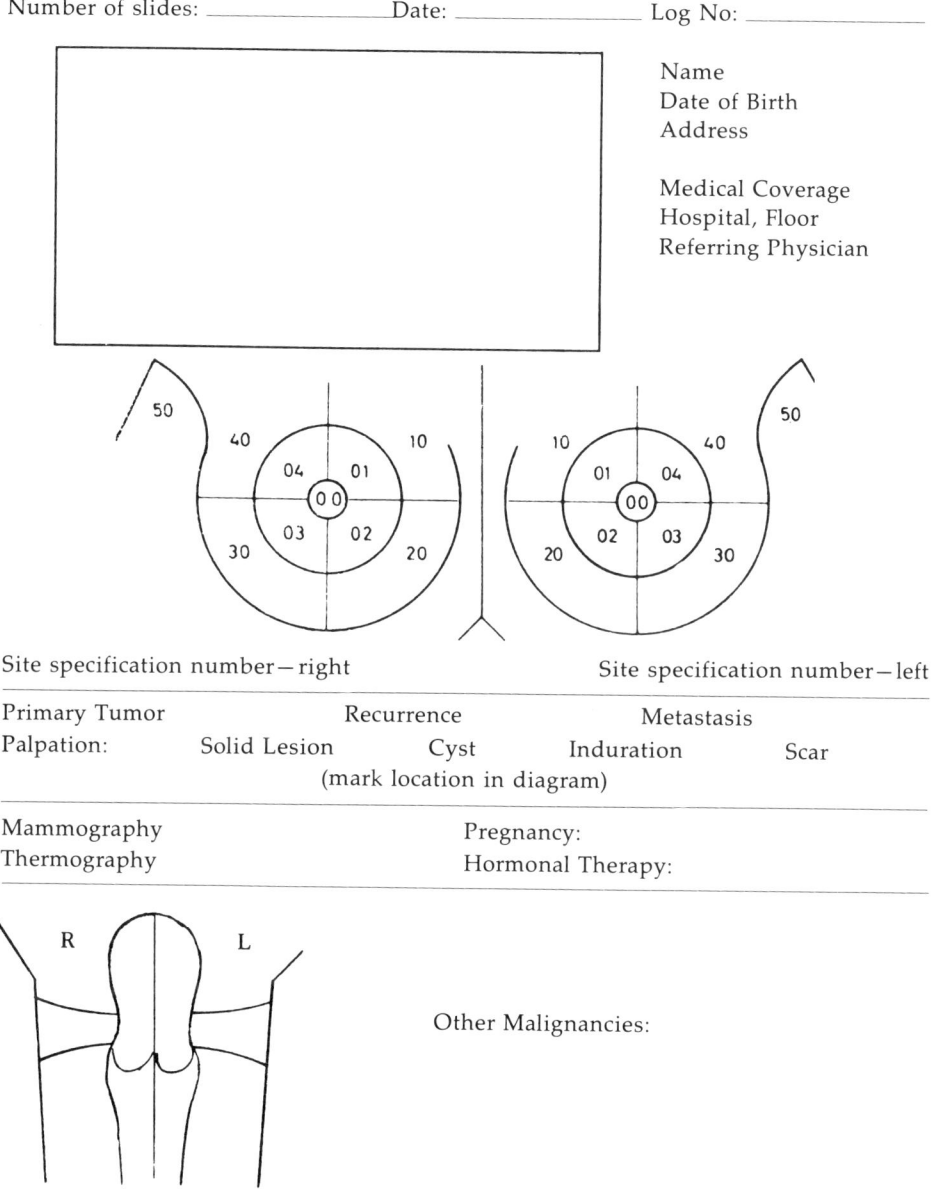

Name
Date of Birth
Address

Medical Coverage
Hospital, Floor
Referring Physician

Site specification number—right

Site specification number—left

Primary Tumor		Recurrence		Metastasis	
Palpation:	Solid Lesion	Cyst	Induration		Scar

(mark location in diagram)

Mammography   Pregnancy:
Thermography   Hormonal Therapy:

Other Malignancies:

Figure 42. Part 1 of the request sheet for needle aspiration of the breast. Information is provided by the requesting physician.

Cytological report:

1) Duct and acinar cells

Number: scarce — moderate — abundant
Degree of Atypia: I   II   III

2) Myoepithelial cells   3) Apocrine cells

4) Foam cells   5) Fat cells

6) Fibrocytes   7) Giant cells

8) Tumor cells

Number: scarce — moderate — abundant
Arrangement: groups — clumps — disseminated
Nuclear size: small — medium — large
Nuclear Shape: pleomorphic — monomorphic
Cytoplasm: rim — naked nuclei
Signet-ring cells — multinucleated cells
Mitosis

9) Background: Mucus — Secretion — Debris — RBC — PMN

Cytological Diagnosis: Unremarkable smear
                             Suspicious smear
                             Tumor cells present

Unsatisfactory specimen

Histology

Date: Signature:

Figure 43. Part 2 of the request sheet for needle aspiration of the breast. This portion is kept on file in our department.

Hospital of the
Johann Wolfgang Goethe-University
Department of Obstetrics and Gynecology
Division of Clinical Cytology

Number of slides:_____   Date:_____   Log No._____

Name
Date of Birth
Address

Medical Coverage
Hospital, Floor
Referring Physician

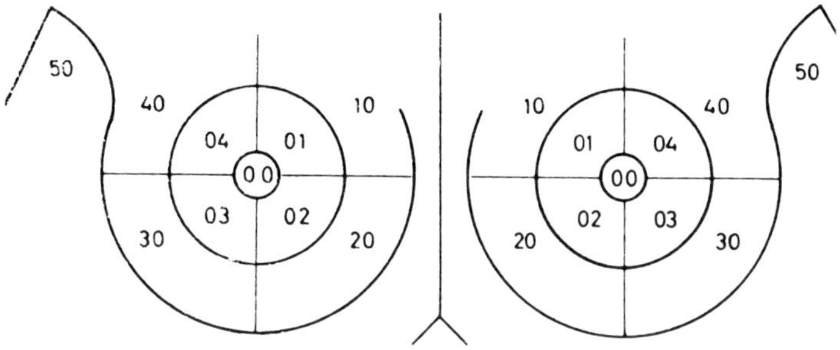

Site specification number—right   Site specification number—left

Cytological Diagnosis: Unremarkable smear
                                  Suspicious smear
                                  Tumor cells present

Unsatisfactory specimen

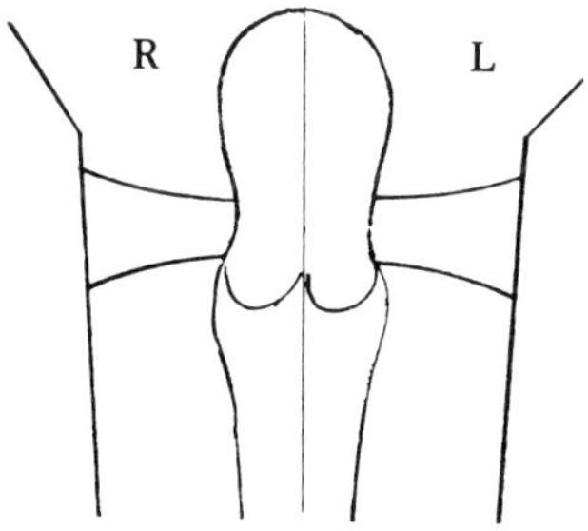

Date: Signature:

Figure 44.   Part 3 of the request sheet for needle aspiration of the breast. This portion
is sent back to the physician with the final diagnosis.

8. The cytological presentation of benign and malignant lesions of the breast

In the presence of adequate and representative material, the cytopathologist is generally able to correctly identify a lesion of the breast as benign or malignant. Additional and more specific information can be provided for a number of lesions.

Cysts and inflammatory lesions may be diagnosed accurately if the clinical findings are known. Approximately 70 per cent of all fibroadenomas show a characteristic appearance of epithelial cells in the form of finger- or antler-like tubules in a richly cellular specimen. Marked cellularity in combination with total dissociation of the cells is typical of infiltrating lobular carcinoma. In aspirates of intraductal carcinoma, tubular arrangements of malignant cells may be observed. Large amounts of mucus containing tumor cells suggest the presence of mucinous carcinoma. Paget's disease of the breast should be suspected if tumor cells with large nuclei and a coarse chromatin pattern are encountered in material obtained from an erosive lesion of the nipple.

In general, all the clinical information, including history, clinical presentation, and results of mammography and thermography, should be considered in each case and correlated, whenever possible, with the evidence found in the cytological specimen.

8.1. The cytological presentation of benign lesions

8.1.1. Cytology and clinical management of large cysts

Cytological evaluation and clinical management of large cysts are closely related, since aspiration represents not only a diagnostic procedure but also final therapy in the majority of cases. Therefore, in dealing with this subject, cytological as well as clinical principles have to be considered. Radiologically, cysts present as well-demarcated round lesions and cannot always be differentiated from other conditions such as fibroma, papilloma, lipoma, hemangioma, adenoma, abscess formation, metastases, or even primary malignant lesions of the breast.[29, 76, 102, 120] Therefore, the aspiration of all well-demarcated lesions of the breast was proposed by Leborgne[67] and Gros[46] as an additional procedure. According to Haagensen[51] five different mechanisms of cyst formation exist: (1) cysts developing in dilated ducts, (2) cysts containing inspissated milk (galactoceles), (3) cysts evolving in duct ectasia, (4) cysts resulting from traumatic fat necrosis, and (5) cysts associated with intraductal papilloma.

Large cysts are usually lined with a single-layered flattened epithelium. Occasionally, the epithelium is missing. In these instances, the connective tissue itself forms the lining of the cyst. The cyst fluid is usually amber-colored, although occasionally it may be greenish-grey, bloody, or brown, owing to the presence of hemosiderin. Usually, only a few cells are found in cyst fluid, most being foam cells. Less often, groups of flattened epithelial cells or apocrine cells may be present.[42, 43, 85] In addition, leukocytes and multinucleated giant cells may be observed (Figs. 23 and 46).

Aspiration of the fluid is the treatment of choice in most institutions concerned with the diagnosis and therapy of lesions of the breast. Surgical excision is avoided if possible.[31, 44, 54, 57, 62, 81, 102] Recurrence is seen in no more than 5 per cent of cases.[31, 57]

Three additional procedures are recommended by several authors for the proper management of large cysts:

1. Cytological examination of the aspirate to rule out the presence of malignant cells.[4, 57, 62, 81, 102, 119]

2. Careful palpation after aspiration with the needle in place in order to rule out residual masses that were previously masked by the cyst and remained undetected.[89, 120]

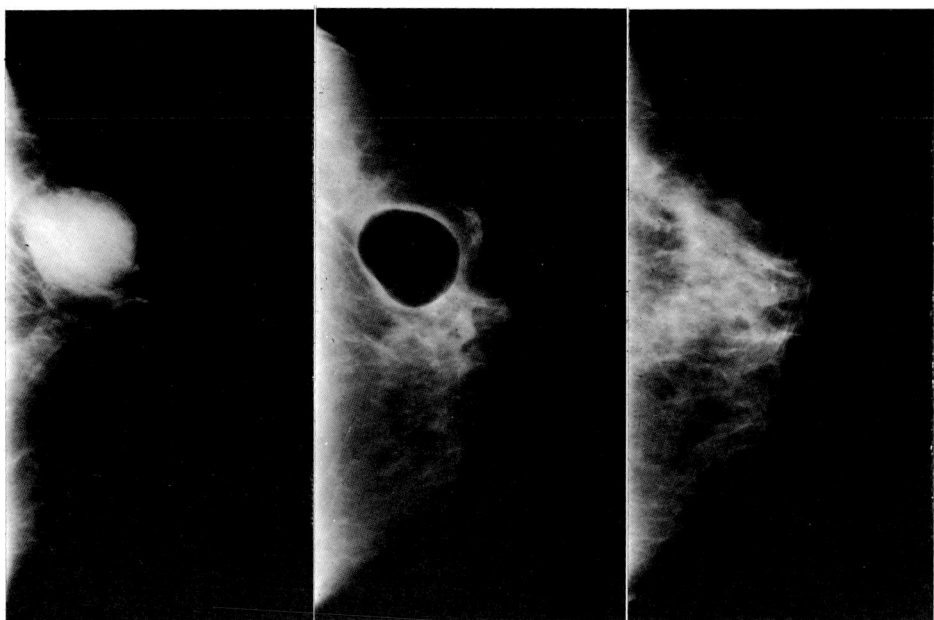

Figure 45.   Aspiration of a large cyst of the breast. (a) Well-demarcated lesion demonstrated by mammography. (b) Aspiration of the fluid and subsequent filling with air (pneumocystography). (c) Control after six weeks. (Courtesy of Professor Dr. Glätzner)

3. Pneumocystography (anteroposterior and lateral) to rule out intracystic carcinoma, wall infiltration, and lesions located in the vicinity of the cyst (Fig. 45).[4, 15, 28, 39, 50, 76]

Carcinoma arising in a cyst is very rare,[123] as indicated in the following table.

TABLE 1. Prevalence of Intracystic Carcinoma

SOURCE	SAMPLE	PERCENTAGE
Gatchell et al.[36]	9000 carcinomas of the breast	0.5
Rosemond et al.[88]	3000 cysts	0.1
Barnes[2]	2000 cysts	0.05

In our own study, 532 large cysts were found among 2778 needle aspirations of the breast. The largest cysts contained up to 50 ml of fluid, the smallest ones that could be palpated and aspirated, 0.5 ml. A cytological examination of the fluid was performed in all cases. The use of a cytocentrifuge greatly facilitates the cytological examination by concentrating the few cellular elements in a small area of the slide (for technical details, see page 12). The cytological examination was unremarkable in 520 cases. In 11 cases, the cytological picture was suspicious. Histological examination of subsequently obtained biopsy material revealed no sign of disease. In one case, malignant cells were found in the fluid (Fig. 47). Histological examination showed a carcinoma in the vicinity of the cyst that had infiltrated into the cyst wall, producing a grossly hemorrhagic fluid.

A major advantage of the treatment of large cysts by aspiration is that surgical exploration and thus hospitalization and anesthetization are unnecessary. Furthermore, fibrosis and scar formation are avoided, thereby facilitating radiological follow-up. However, in order to make this type of management an absolutely safe procedure, the additional examinations mentioned earlier should be performed.

An excisional biopsy should be done in the presence of:

1. Abnormal results of pneumocystography, e.g., induration or irregularity of the cyst wall, papillary structures in the lumen, or multiloculation.
2. Suspicious or positive results of the cytological examination.
3. Hemorrhagic fluid, even with negative cytological results.[59, 89]
4. Masses remaining after fluid aspiration.[44, 59, 89]
5. Third recurrence of fluid accumulation.[43]

8.1.2. Benign tumors of the breast

The benign character of a lesion usually is easily recognized in needle aspirates of the breast. Additional evidence provided by mammography and thermography is often available to corroborate the cytological diagnosis. However, it is not possible on the basis of cytological findings alone to specify the type of underlying tumor.

Linsk and his associates[70] compared the cytological presentation of 210 aspirates obtained from histologically confirmed cases of fibroadenoma with the same number of smears obtained from fibrocystic disease. In this study, smears from fibroadenoma were found to have a higher degree of cellularity. In addition, only in fibroadenoma were fragments of stromal tissue present. However, the cellular elements are identical in both lesions, i.e., duct cells, acinar cells, apocrine cells, foam cells, and naked bipolar nuclei. Therefore, a clear-cut separation of these two entities is not possible on the basis of cytological examination.

In the differential diagnosis of benign lesions of the breast and carcinoma, degree of cellularity and cellular dissociation are the main criteria. Aspirates obtained from carcinoma of the breast usually show rich cellularity and total dissociation of the malignant cells. In benign lesions of the breast, as in fibrocystic disease, the smear usually contains few cells. In some cases of fibroadenoma, rich cellularity may be present. In these instances, the cells are tightly packed together, forming finger- or antler-like structures. In addition, these clusters show two layers of cells or, if present as a monolayer, form a flat sheet of cells with preservation of nuclear polarity (Figs. 8, 9, 11, 48, and 49).

Some cases of fibroadenoma and fibrocystic disease present with a marked degree of nuclear pleomorphism and enlargement.[27, 119] In order to investigate the correlation between the degree of atypia in the cytological smear and the amount of epithelial proliferation in the histological section, smears from 60 histologically confirmed cases of fibroadenoma and fibrocystic disease were reexamined. As described earlier (see page 61), a classification into three degrees of cytological atypia was made. In the vast majority of cases, no correlation between cytological and histological findings could be demonstrated. Similar results were reported by Finsterer and his associates.[30] These authors examined the smears of 97 cases of mastopathy and found no correlation between cytological atypia and degree of epithelial proliferation in the tissue.

The presence of nuclear enlargement and pleomorphism in some benign lesions creates additional problems in ruling out malignancy. In these cases, the entire specimen should be screened carefully in order to assess the numerical relation between benign and malignant-appearing cells. Naked bipolar nuclei are of special importance to the differential diagnosis in this situation[70] because their presence makes a carcinoma of the breast highly unlikely. However, it should be kept in mind that naked bipolar nuclei may be aspirated accidentally as the needle passes through normal tissue to the area of malignant disease. In our series, occasional naked bipolar nuclei were observed in 10 per cent of cases of carcinoma of the breast. Geier and his colleagues[38] reported similar results.

8.1.3. The cytological presentation of mastitis

An abundance of so-called inflammatory cells, i.e., leukocytes, granulocytes, lymphocytes, monocytes, and histiocytes, is the hallmark of aspirates obtained from mastitis. In addition, foam cells and multinucleated giant cells are often present. In advanced stages of inflammation, fine needle aspiration smears have a dirty appearance, owing to cellular debris and amorphous material in the background. Duct cells often show nuclear enlargement (Fig. 57), but inflammatory carcinoma can usually be ruled out (Figs. 68 and 69). However, atypical histiocytes may occasionally present more problems than duct cells for the differential diagnosis. Nuclei of histiocytes may increase markedly in size and vary considerably in shape. The presence of finely vacuolated, delicate cytoplasm with ill-defined borders is the main criterion for correct identification. Furthermore, in contrast to malignant cells, histiocytes possess a smooth nuclear membrane without irregularities of contour (Figs. 56 and 58). In this situation, it is extremely important to study the entire specimen unhurriedly and to examine the nuclear membrane carefully at higher magnification ($\times$ 400).

8.1.4. The cytological presentation of fat necrosis

Fat necrosis is a condition that should always be kept in mind when dealing with lesions of the breast. A history of trauma can usually be elicited and is an important clue to the correct diagnosis. In the cytological smear, solitary fat cells, clusters of fat cells, polymorphonuclear leukocytes, giant cells, and a fair number of histiocytes are observed on a background containing amorphous material (Figs. 58, 59, and 60).[42]

8.1.5. The cytological picture of breast aspirates during gestation

Needle aspirates of the breast obtained during pregnancy show marked enlargement of the nuclei and prominent nucleoli resulting from hormonal stimulation and increased secretory activity. However, the nuclear diameter rarely reaches values of 12 μ or greater. Marked cellularity is the rule, with the predominance of large round nuclei and clear cytoplasm giving the smear a rather monomorphous appearance. The differential diagnosis between these alterations, characteristically observed during gestation, and the so-called acinic cell carcinoma[122] is particularly difficult, since the nuclear appearance is almost identical in both conditions.

Foam cells may show nuclear enlargement and multinucleation in pregnancy. Histiocytes and polymorphonuclear leukocytes are usually increased in number (Figs. 25 and 61).[47, 62]

Figures 46 through 61 appear in the following section.

Figure 46.   Needle aspiration of a large cyst. Large foam cells and a group of apocrine cells. Cytocentrifuge. May-Grünwald-Giemsa × 400.

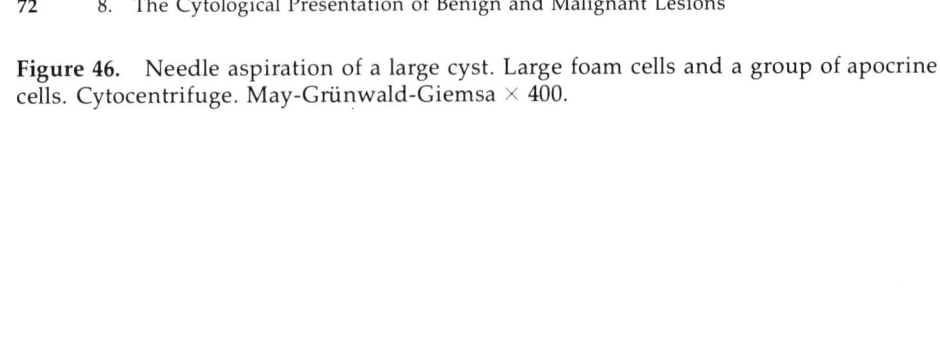

Figure 47.   Needle aspiration of a large cyst. Tumor cells with marked pleomorphism. After aspiration of the fluid, a mass could be palpated in the vicinity of the cyst. Cytocentrifuge. May-Grünwald-Giemsa × 400.

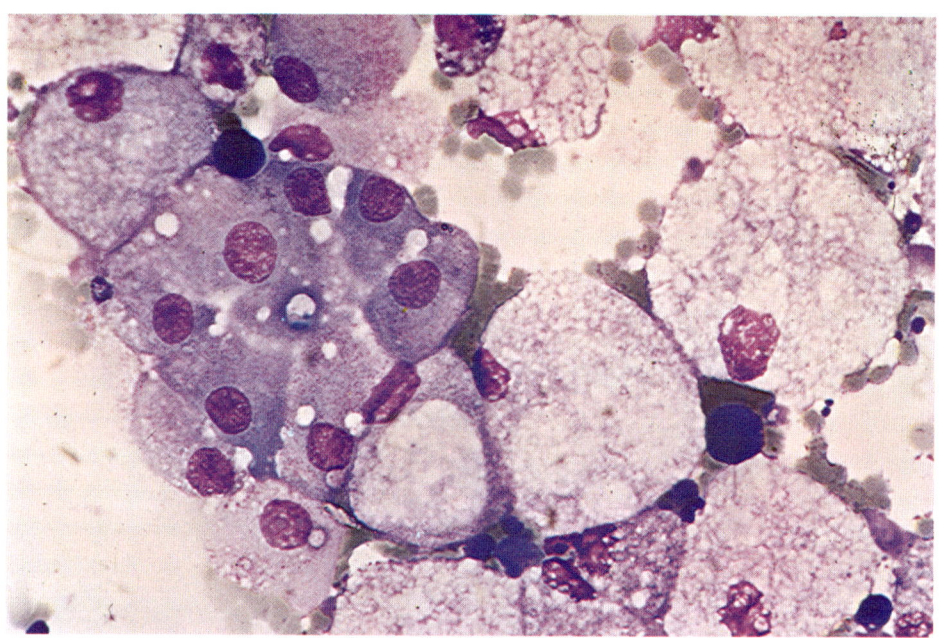

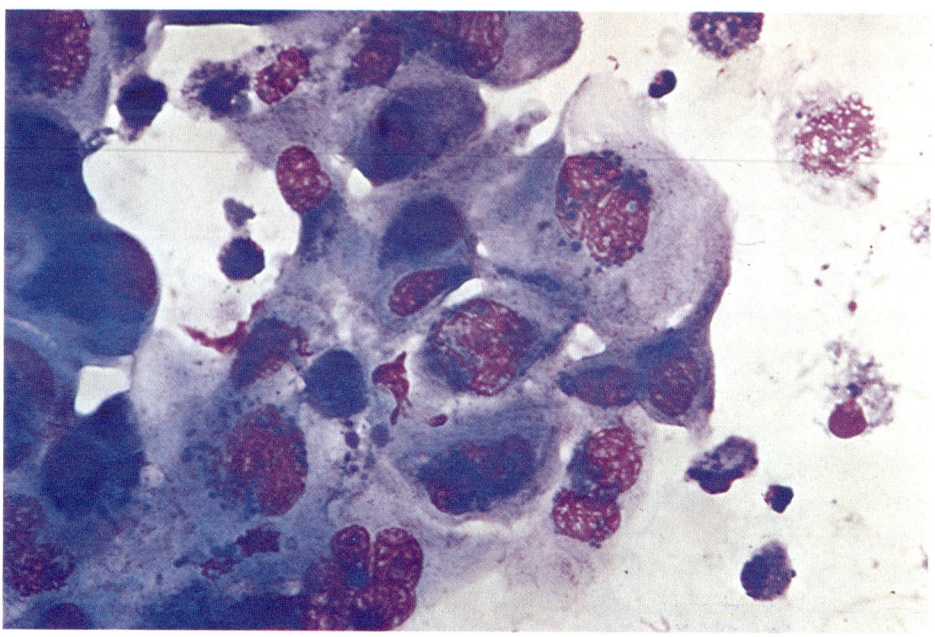

Figure 48. Cluster of branching epithelial cells. Aspirate of fibroadenoma. May-Grünwald-Giemsa × 400.

Figure 49. Cluster of unremarkable duct cells. Typical tubular arrangement with an attached acinus. Aspirate of fibroadenoma. May-Grünwald-Giemsa × 400.

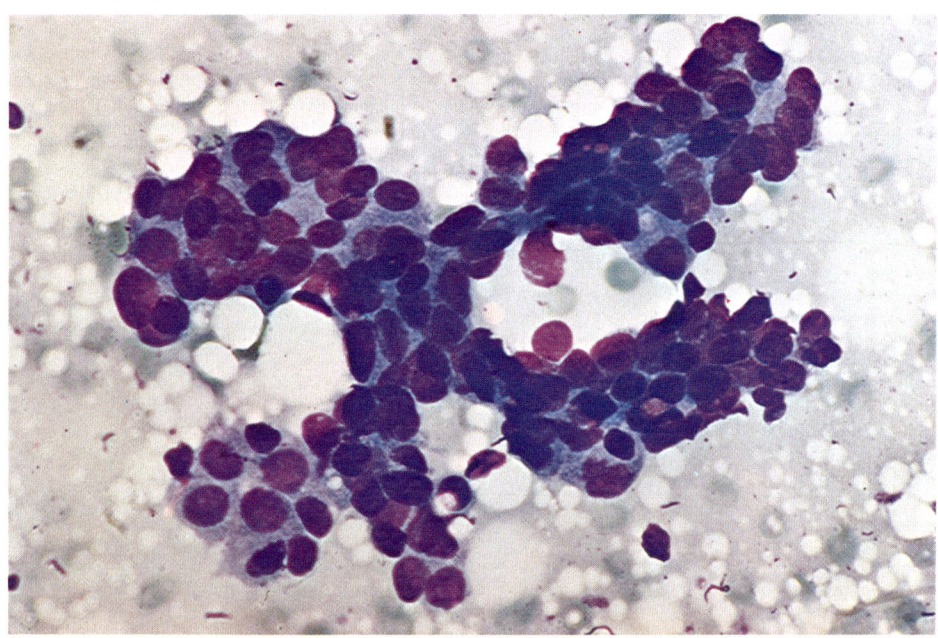

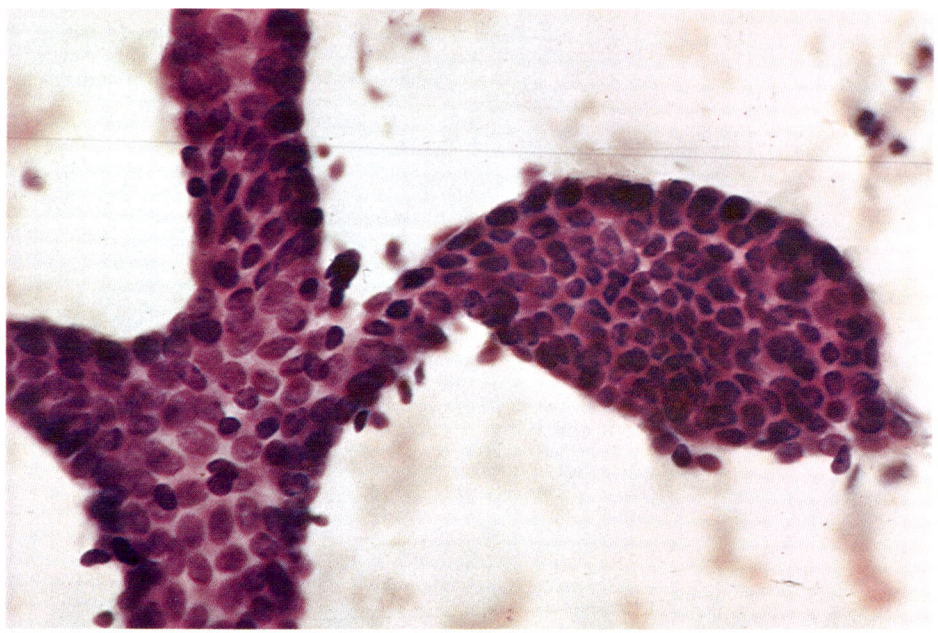

Figure 50.   Duct and acinar cells with some variation in nuclear size and shape that is well within normal limits. Several naked bipolar nuclei at the margins of the cell cluster. Nuclear size can be evaluated easily by comparison with the size of erythrocytes. Aspirate from fibroadenoma. May-Grünwald-Giemsa × 400.

Figure 51.   Duct and acinar cells with variation in size and shape. In comparison with erythrocytes, some of the nuclei appear to be markedly enlarged. Prominent nucleoli. Cytological diagnosis: suspicious smear. Histological examination: fibroadenoma. May-Grünwald-Giemsa × 400.

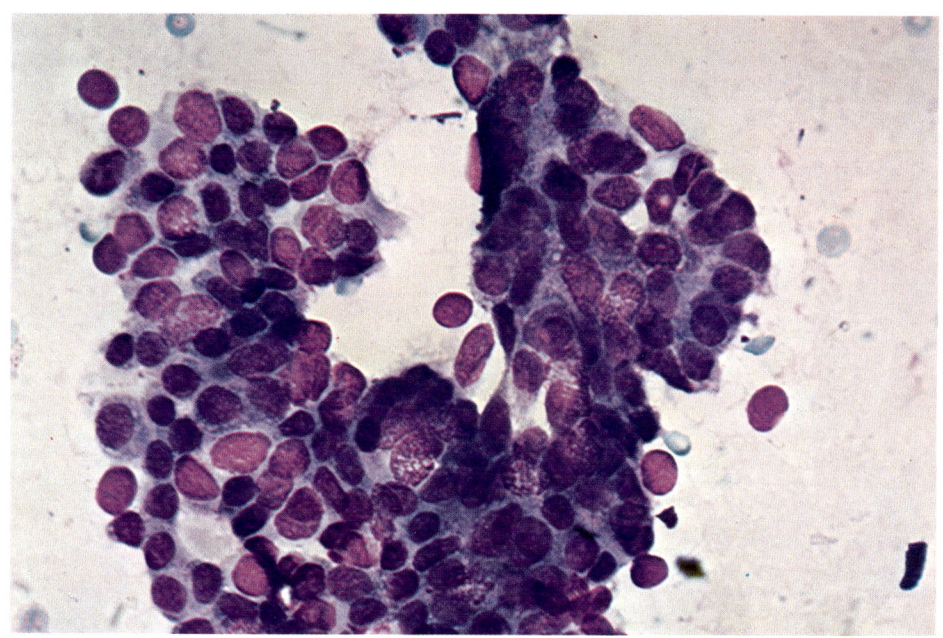

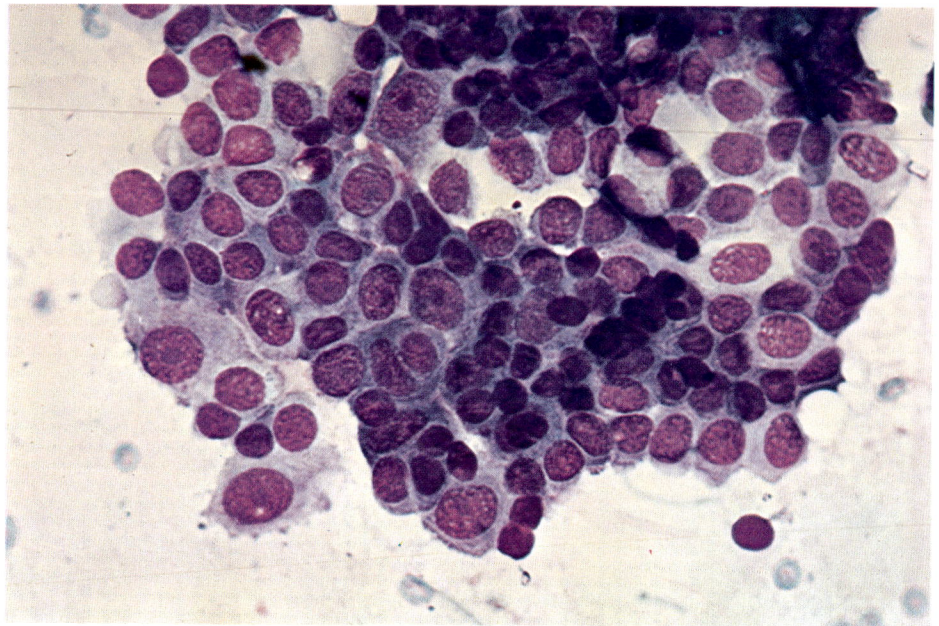

Figure 52. Unremarkable duct cells with small uniform nuclei (lower left corner). Markedly enlarged nuclei with mild pleomorphism (upper right corner). Cytological diagnosis: suspicious smear. Histological examination: intracanalicular fibroadenoma and diffuse fibrocystic disease. May-Grünwald-Giemsa × 400.

Figure 53. Aspirate containing markedly enlarged nuclei with mild pleomorphism. These nuclei were erroneously diagnosed as malignant. Histological examination: proliferating fibroadenoma and proliferating mastopathy. May-Grünwald-Giemsa × 400.

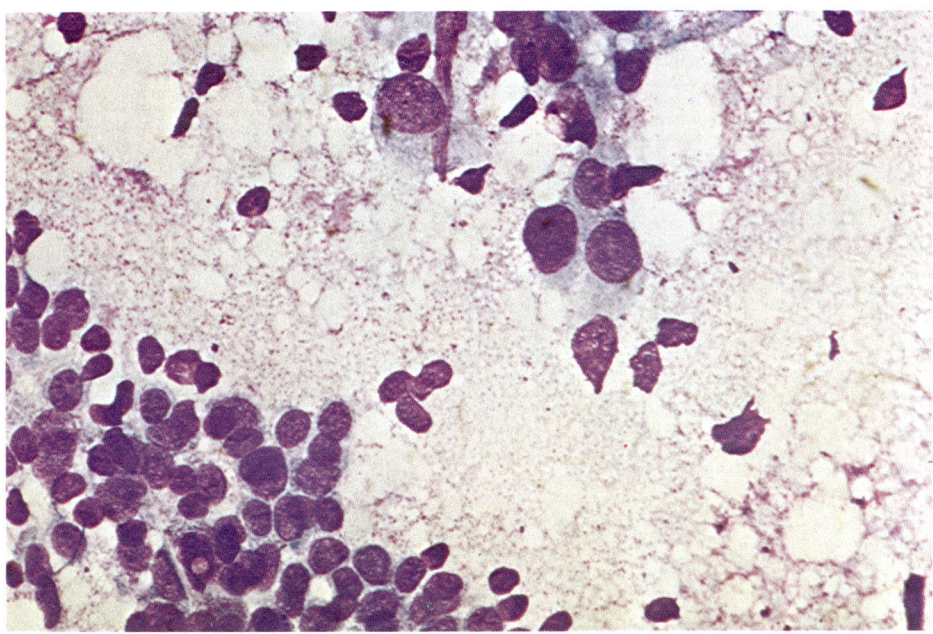

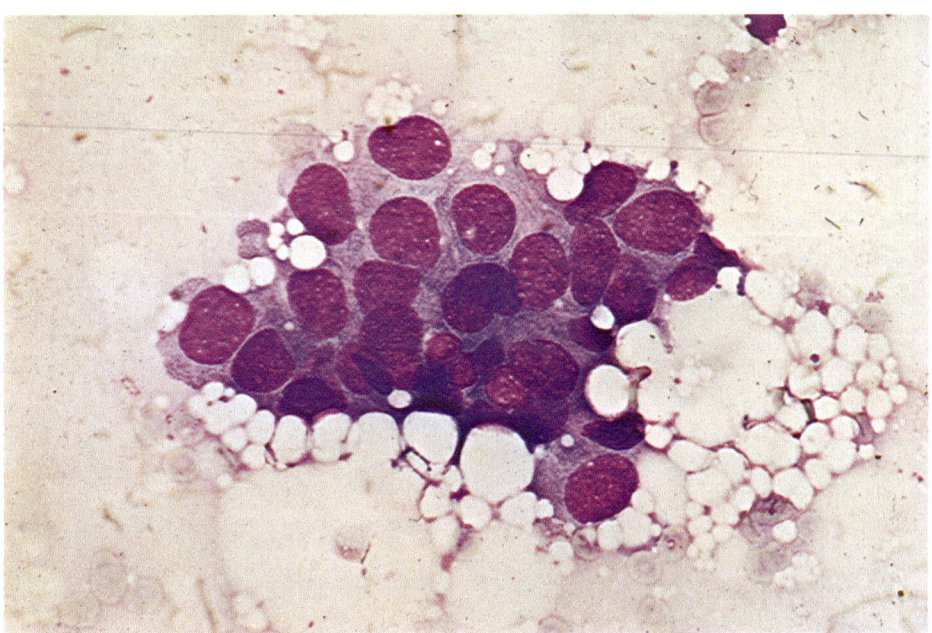

Figure 54. Moderately enlarged nuclei with marked pleomorphism. These nuclei were considered to be malignant by the cytopathologist. Histological examination: fibroadenoma. May-Grünwald-Giemsa × 400.

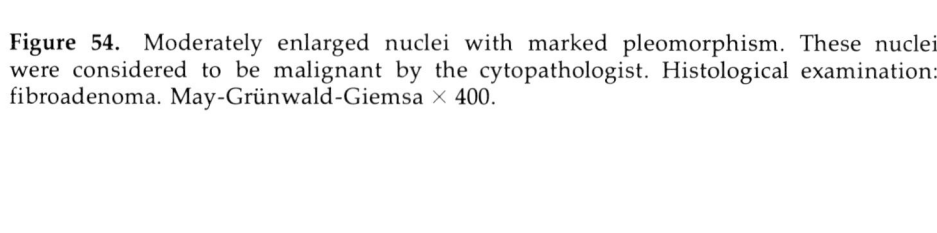

Figure 55. Larger group of unremarkable duct and acinar cells (right). Naked bipolar nuclei, foam cell, and erythrocytes (left) in a generally poorly cellular aspirate. Histological examination: fibrocystic disease. May-Grünwald-Giemsa × 400.

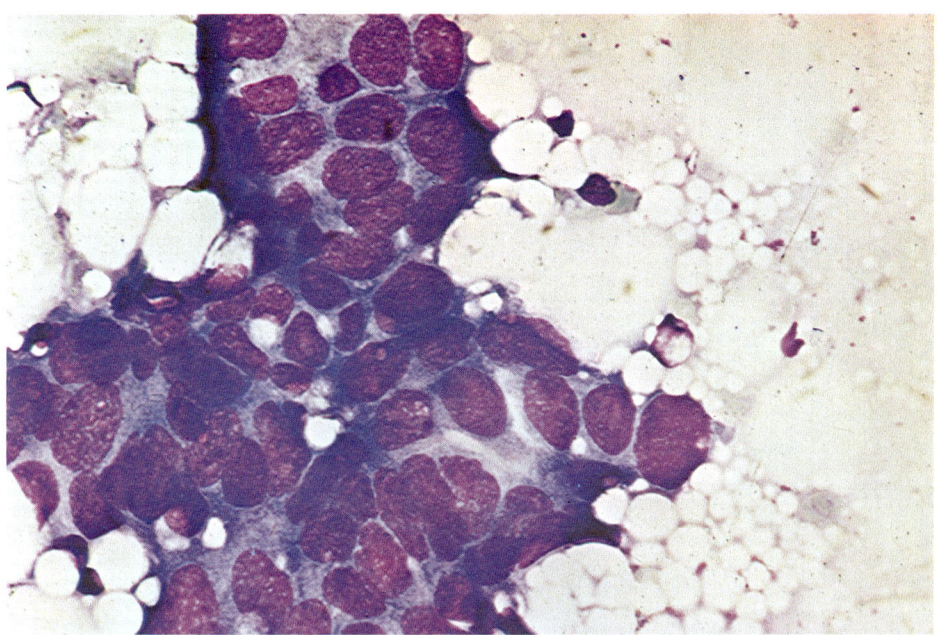

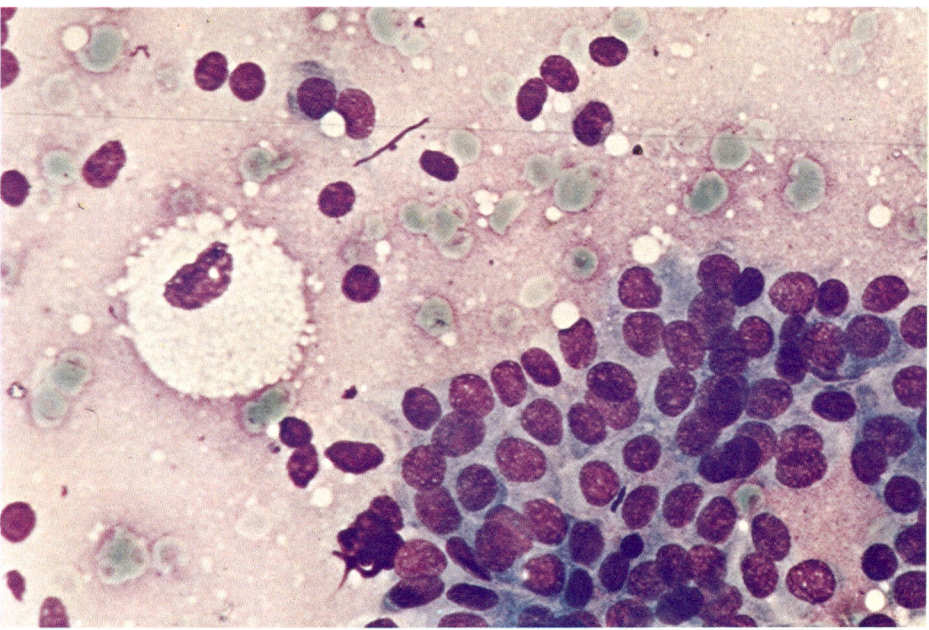

Figure 56.   Aspiration of acute mastitis. Large histiocytes with finely vacuolated, ill-defined cytoplasm and numerous polymorphonuclear leukocytes. May-Grünwald-Giemsa × 400.

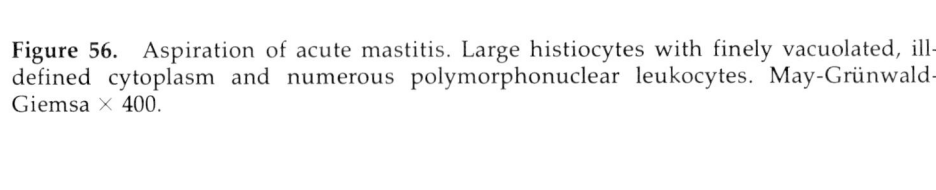

Figure 57.   Aspiration of acute mastitis. Epithelial cells with markedly enlarged nuclei, prominent nucleoli, coarsely granular chromatin, but generally monomorphous appearance. Numerous polymorphonuclear leukocytes. May-Grünwald-Giemsa × 400.

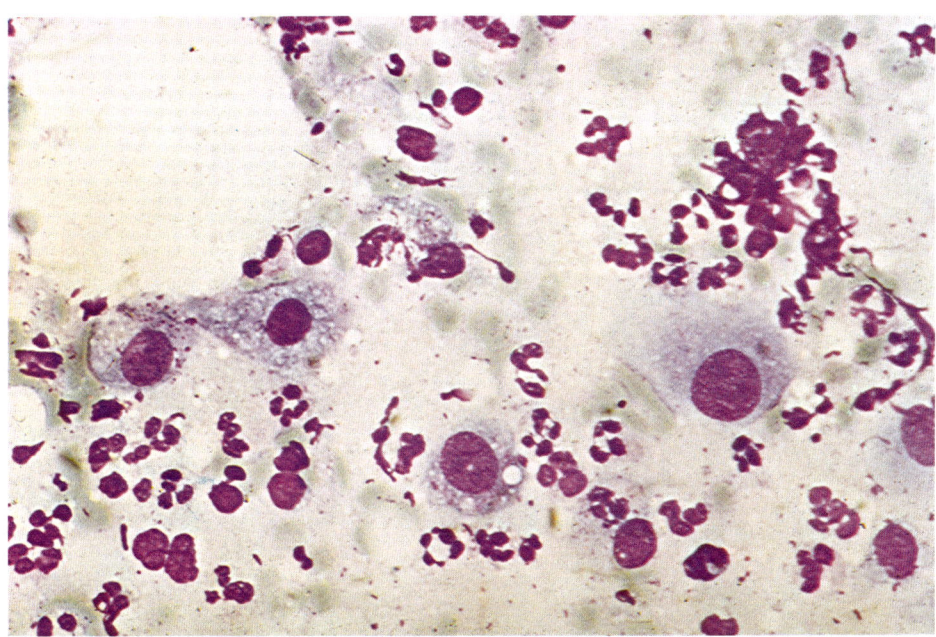

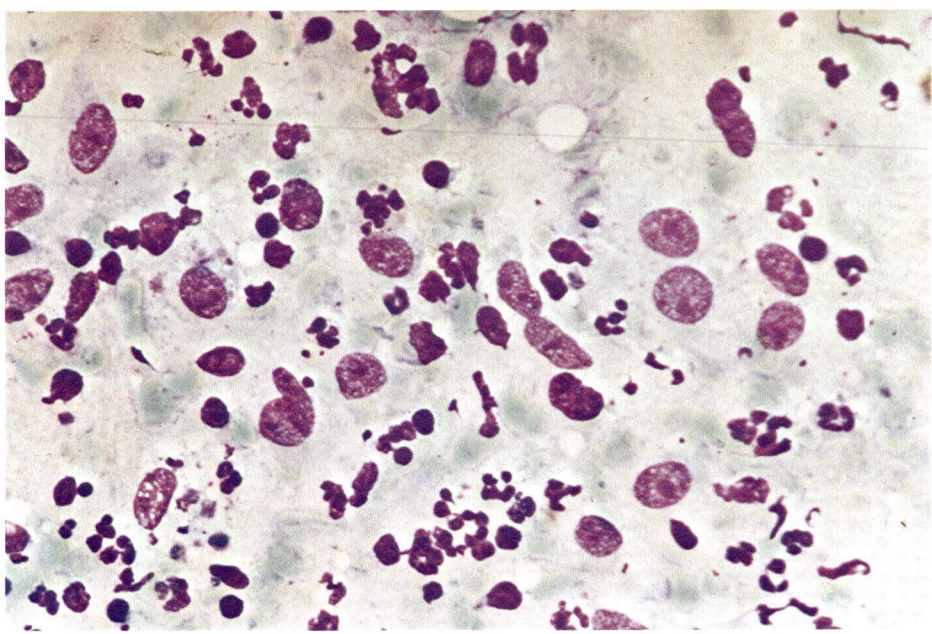

Figure 58. Aspiration of fat necrosis of the breast. Collapsed cytoplasmic membranes of fat cells (center). Histiocytes (top) with vacuolated cytoplasm. Polymorphonuclear leukocytes. May-Grünwald-Giemsa × 400.

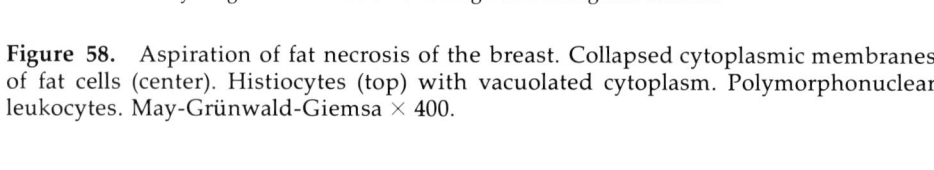

Figure 59. Aspiration of fat necrosis of the breast. Histiocytes with oval nuclei and vacuolated cytoplasm (lipophages) surrounding cytoplasmic fragments of fat cells. May-Grünwald-Giemsa × 400.

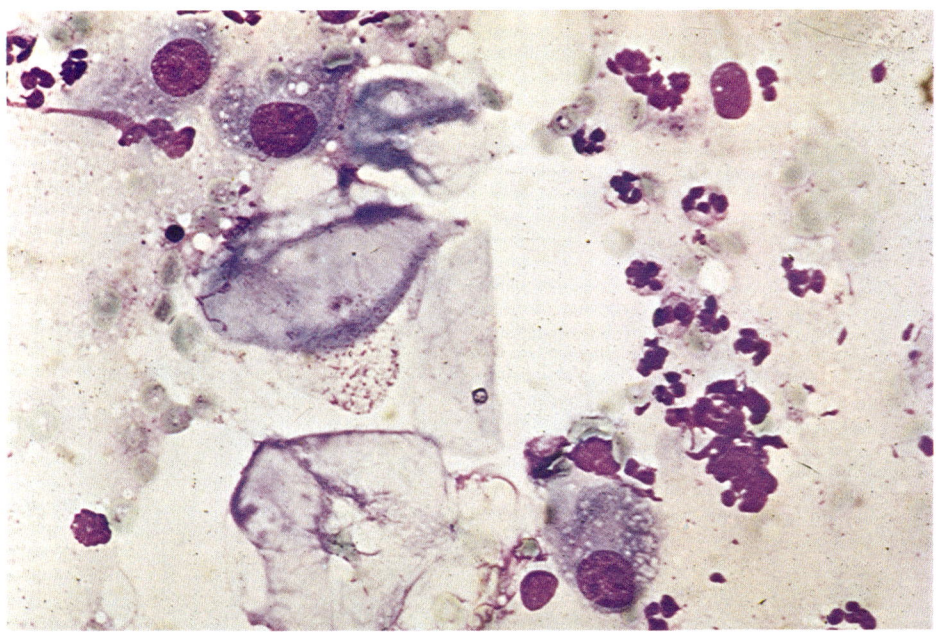

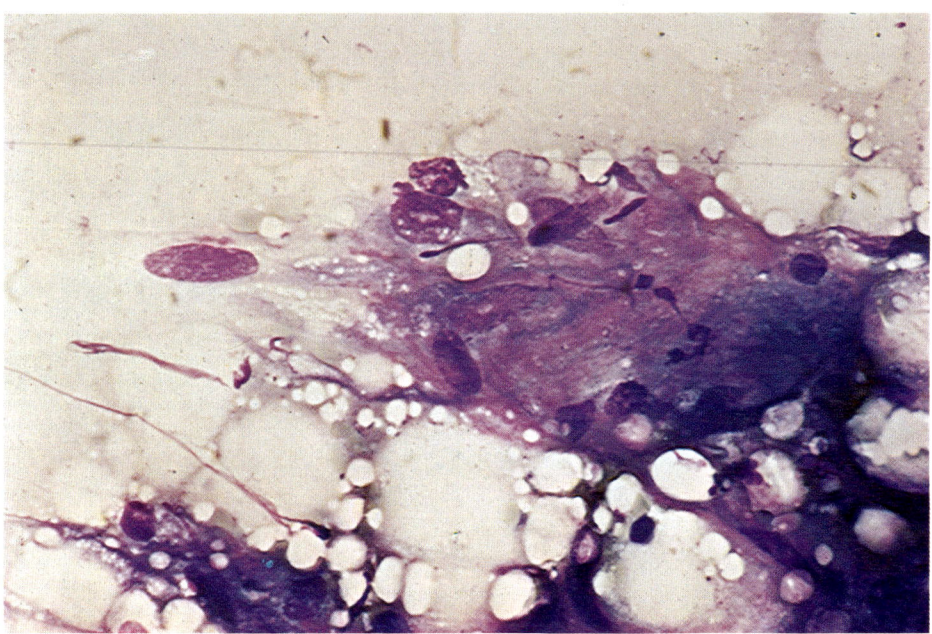

Figure 60.   Aspiration of fat necrosis of the breast. Histiocytes with large, pleomorphic nuclei and vacuolated, ill-defined cytoplasm. The erroneous diagnosis of this type of nuclei as malignant is one of the classical mistakes made in fine needle aspiration of the breast. May-Grünwald-Giemsa × 400.

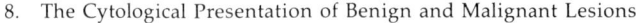

Figure 61.   Needle aspiration of the breast during pregnancy (32 weeks). Epithelial cells with markedly enlarged nuclei, coarse chromatin pattern, and occasional nucleoli. May-Grünwald-Giemsa × 400.

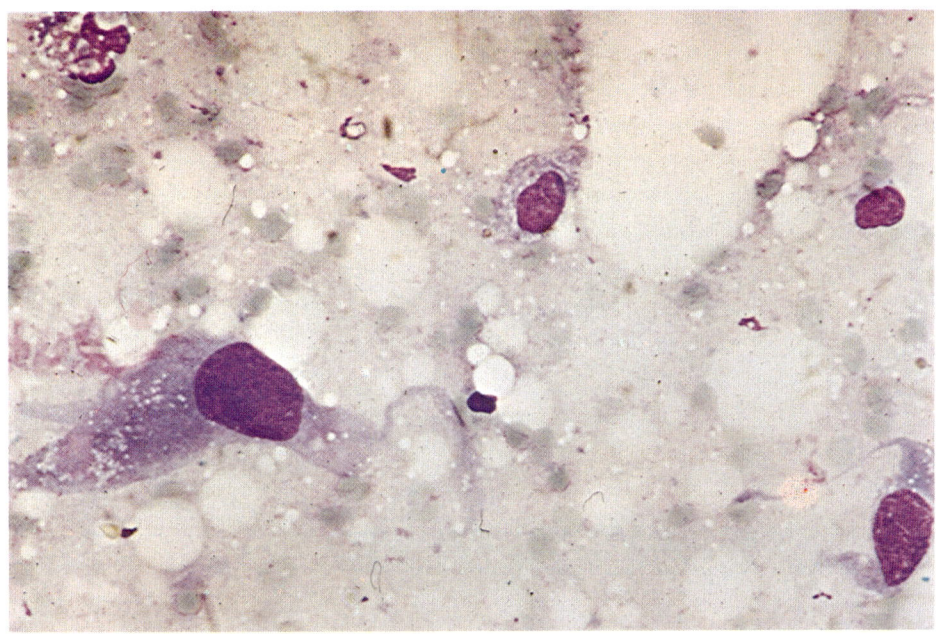

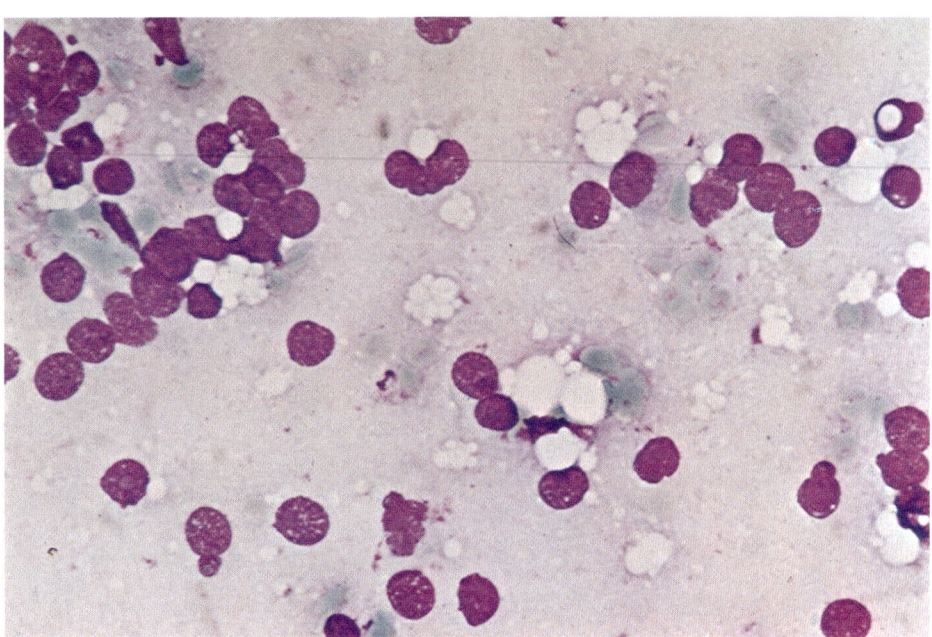

8.2. The cytological presentation of malignant lesions

8.2.1. Duct cell carcinoma and lobular carcinoma

As previously mentioned, the interpretation of fine needle aspirations of the breast is limited to recognition of the benign or malignant potential of the cells. In order to specify the type of lesion present, the examination of tissue sections is necessary. Certain rare tumor types may be recognized in cytological material because of their specific features, e.g., mucinous carcinoma, inflammatory carcinoma, and Paget's disease of the breast.

In differentiating duct cell carcinoma from lobular carcinoma, Zajicek was able to demonstrate that certain criteria were helpful in the cytological identification of these two tumor types. In aspirates from well-differentiated duct cell carcinoma, the malignant cells often formed tubular structures (Fig. 63). In infiltrating lobular carcinoma, the tumor cells were totally dissociated, with an occasional rim of well-preserved cytoplasm (Fig. 62).

It should be kept in mind that it is not possible to differentiate between invasive and *in situ* lesions on the basis of cytological findings, that is, between infiltrating duct cell carcinoma and infiltrating lobular carcinoma on the one hand and intraductal carcinoma and lobular carcinoma *in situ* on the other. In cytological smears, the invasive potential of a tumor is no longer evident, since the cells have been removed from their natural environment and placed in a single cell suspension. The relationship of malignant cells to surrounding tissue therefore can no longer be evaluated. Occasionally, atypical or frankly malignant cells may be present, which are later shown histologically to correspond to an *in situ* carcinoma. This is an incidental finding, because *in situ* lesions are not palpable.

The degree of differentiation of a neoplastic process can normally be assessed readily in cytological specimens. A classification into three degrees of differentiation was proposed by Zajdela and his colleagues[116] and other investigators.[18, 112]

> *Grade I*: Well-differentiated carcinoma. Marked uniformity of the nuclei, rim of cytoplasm present, and glandular differentiation prevailing.
> *Grade II*: Moderately well-differentiated carcinoma. Mixture of gland-forming and anaplastic-appearing cells.
> *Grade III*: Poorly-differentiated carcinoma. Total dissociation of the nuclei, pleomorphism, and absence of cytoplasm.

Thus, duct cell carcinoma presents in its poorly differentiated form with total dissociation of the nuclei. Further information on cytological classification of breast carcinoma is provided by Zajicek.[122]

8.2.2. Paget's disease of the breast

Paget's disease of the breast is considered today to be an unusual manifestation of duct cell carcinoma, representing metastatic involvement of the epidermis of the nipple and areola.[5] In our study of 333 carcinomas of the breast, six cases of Paget's disease of the breast were encountered. In two of these cases, the diagnosis was established cytologically. Both patients had received unsuccessful dermatological treatment for erosion of the nipple for three and one-half and two years, respectively. A scraping of the erosive lesion revealed malignant cells. In one patient, a palpable lesion in the retromamillary area was aspirated and neoplastic cells were obtained.

Paget's disease of the breast can usually be diagnosed easily by examination of scrapings of the nipple erosion. Prior to nipple changes, the lesion is only rarely detected.[23, 51] The malignant cells in these specimens, so-called Paget cells, are exceptionally large, reaching on the average a nuclear diameter of 16 μ. Despite a rather monomorphous picture, the cells are easily identified as malignant (Figs. 64 and 65).[80] In addition, atypical squamous cells with markedly enlarged, hyperchromatic nuclei, anucleated squames, and cellular debris are often observed in touch preparations or scrapings from Paget's disease of the breast (Fig. 64).[99]

8.2.3. Mucinous carcinoma

Mucinous carcinoma is a rare lesion. In our series of 333 carcinomas of the breast, only four mucinous carcinomas were observed. Characteristically, large amounts of mucus are aspirated, staining reddish-purple with May-Grünwald-Giemsa and more eosinophilic with the Papanicolaou technique. Groups of malignant cells with occasional gland formation are found in the mucus.[18, 72] The nuclei are small and uniform (Figs. 66 and 67). Signet-ring cells were reported to be a characteristic feature of mucinous carcinoma. In our material, signet-ring cell formation was observed in a variety of breast tumors. However, none of the mucinous carcinomas of our series showed this feature.

Fibroadenoma with myxoid degeneration may be diagnosed erroneously as mucinous carcinoma. Naked bipolar nuclei are usually found in adenofibroma, suggesting the presence of a benign lesion.

8.2.4. Inflammatory carcinoma

Inflammatory carcinoma is encountered even less often than mucinous carcinoma. According to Haagensen,[51] clinical presentation in the form of inflammatory carcinoma is not limited to a single histological type of breast cancer. Rather, all known types of carcinoma of the breast may be present in this fashion. Moreover, large numbers of lymphocytes, which one might expect to see in view of the gross presentation, are not a characteristic finding in histological sections of inflammatory carcinoma. In Haagensen's series of 59 inflammatory carcinomas, an unusual degree of lymphocytic infiltration was seen in only six cases.

The clinical picture in inflammatory carcinoma — redness, swelling, induration, and pain—is identical to that of acute mastitis and abscess formation. Therefore, surgical incision is often performed initially. If no pus is obtained, fine needle aspiration is the method of choice to verify the malignant nature of the lesion. The two cases of inflammatory carcinoma in our department yielded numerous malignant cells on aspiration. Numerous polymorphonuclear leukocytes, lymphocytes, and histiocytes were present in the background (Figs. 68 and 70). Acute mastitis could be ruled out easily (Fig. 69) in view of the marked pleomorphism and nuclear enlargement of the malignant cells.

Local extension to the skin and subsequent ulceration do occasionally occur in carcinoma of the breast. Clinically, this condition may superficially resemble inflammatory carcinoma. However, these conditions represent two distinct entities and therefore should be differentiated. Cytologically, the aspirate from ulcerating carcinoma with skin involvement contains numerous malignant cells and a variable amount of inflammation and necrosis (Fig. 71).

8.2.5. Special types

Apocrine carcinoma[122]

Apocrine carcinoma is also known as sweat gland carcinoma or oncocytic carcinoma. Characteristically, the malignant cells contain abundant eosinophilic cytoplasm and have a prominent cell membrane. In well-differentiated forms of apocrine carcinoma, a monomorphous picture with round nuclei is observed, which makes it difficult to rule out benign apocrine metaplasia. In poorly differentiated forms, the malignant character of the cells can easily be recognized by their prominent pleomorphism (Figs. 72 and 91).

Acinic cell carcinoma[122]

Fine needle aspirates classified as acinic cell carcinoma are richly cellular and contain cells with round nuclei, prominent nucleoli, and a small rim of cytoplasm. The cell boundaries are ill-defined, and naked nuclei may be present. The appearance of acinic cell carcinoma resembles the cytological picture seen in aspirates from pregnant or lactating patients (Fig. 74). Histologically, infiltrating duct cell carcinoma is often found as an underlying condition.[122]

Fibrosarcoma

Rich cellularity is characteristic of aspirates from fibrosarcoma of the breast. At lower magnification, large pieces of stroma can be identified. At higher magnification, marked nuclear pleomorphism of stromal cells becomes obvious and confirms their identification as malignant fibrocytes (Figs. 76 and 77).

Figures 62 through 77 appear in the following section.

Figure 62. Aspirate from infiltrating lobular carcinoma. Tumor cells with a small rim of cytoplasm. May-Grünwald-Giemsa × 400.

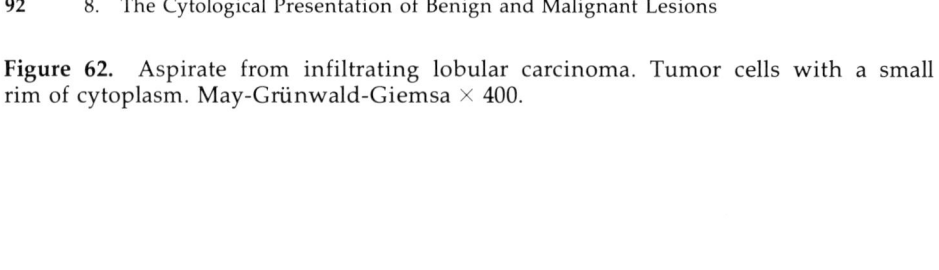

Figure 63. Aspirate from infiltrating duct cell carcinoma. Tubular arrangement of tumor cells. May-Grünwald-Giemsa × 400.

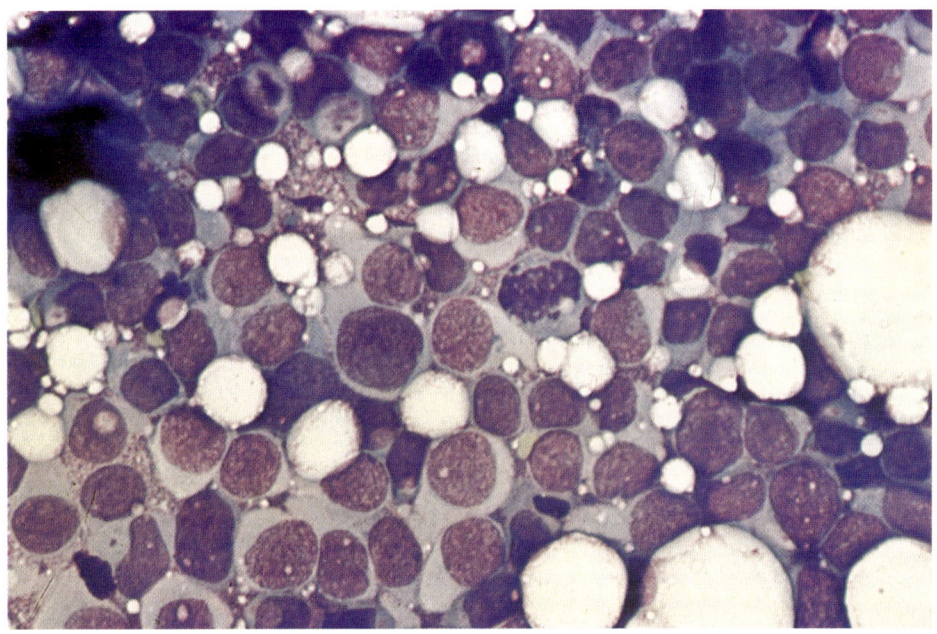

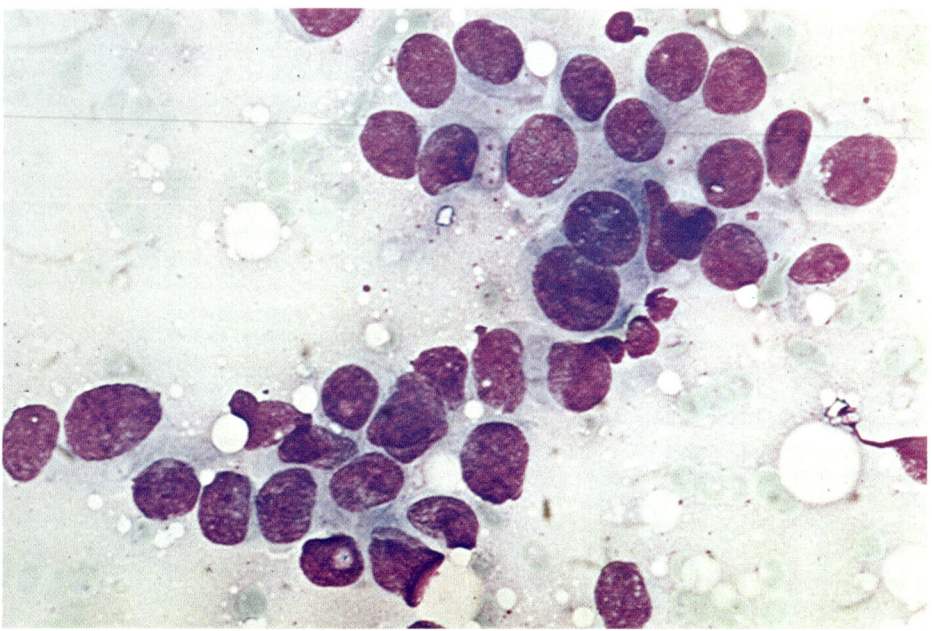

Figure 64.   Smear from an erosion of the nipple. Cluster of malignant cells. Several markedly enlarged nuclei (Paget cells). Anucleated squames. May-Grünwald-Giemsa × 400.

Figure 65.   Aspirate from Paget's disease of the breast. Large tumor cells. May-Grünwald-Giemsa × 400.

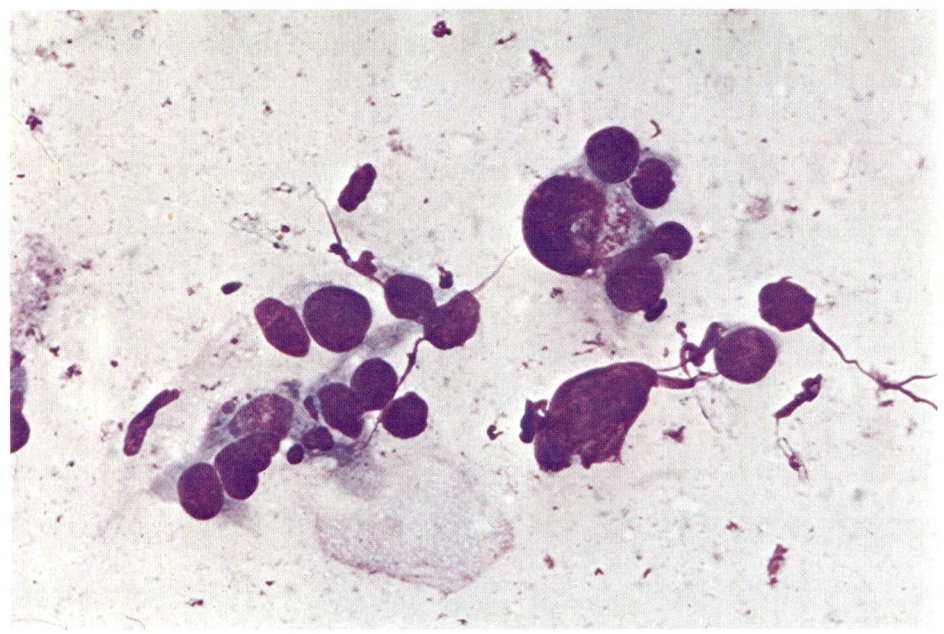

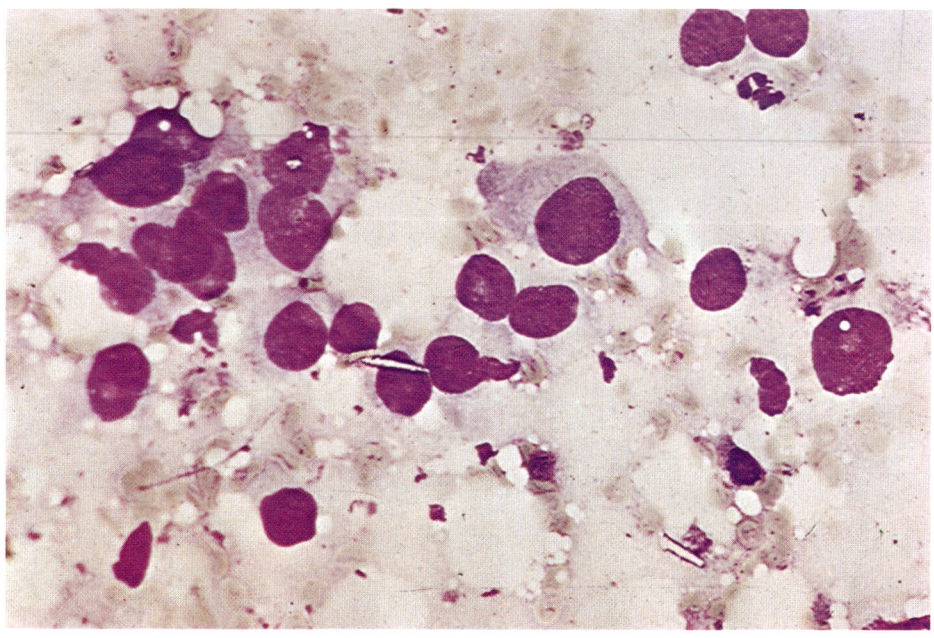

Figure 66.   Aspirate from mucinous carcinoma. Group of monomorphous tumor cells embedded in mucus. May-Grünwald-Giemsa × 400.

Figure 67.   Aspirate from mucinous carcinoma. Groups of monomorphous tumor cells. Papanicolaou × 400.

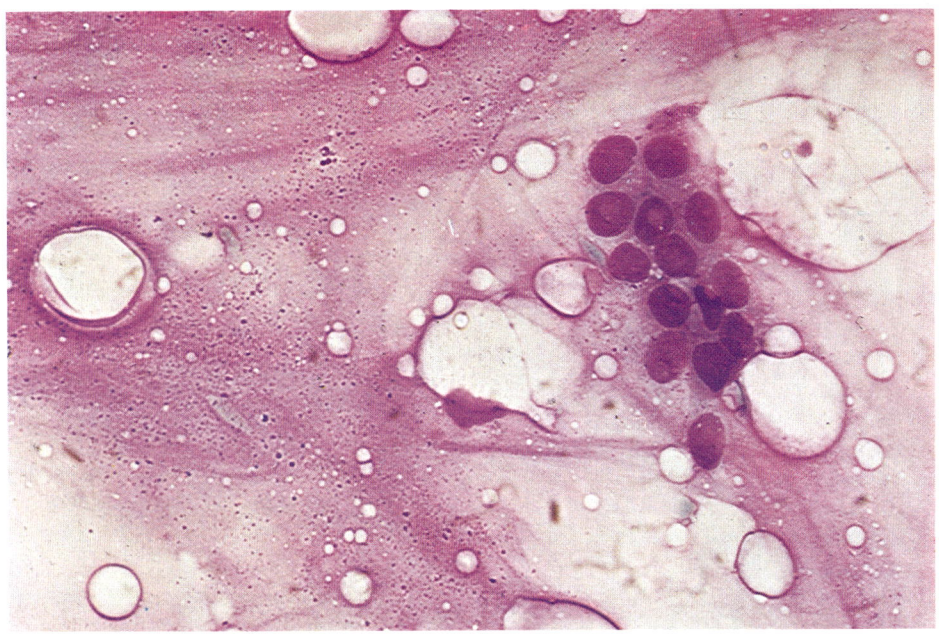

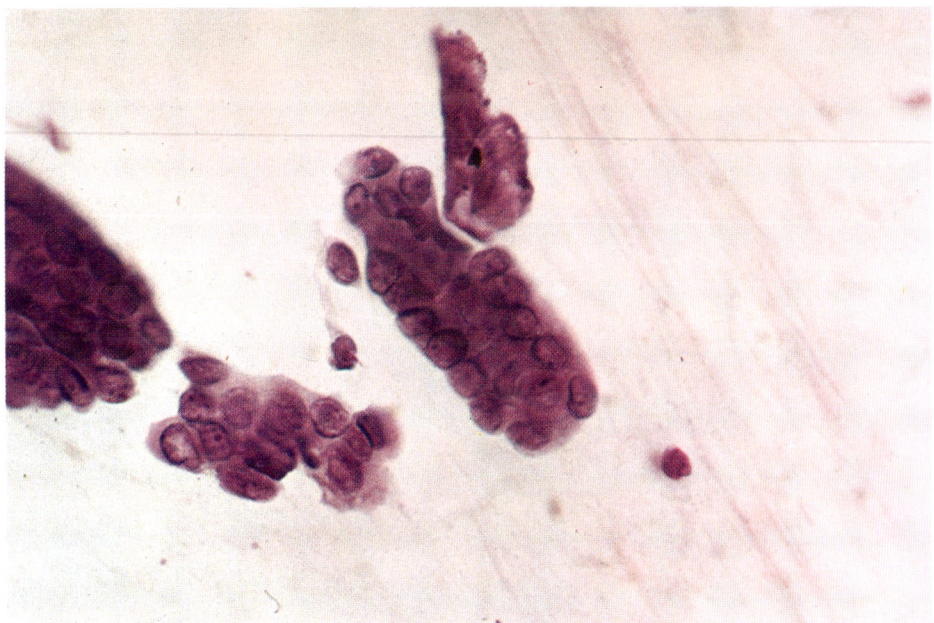

Figure 68.   Aspirate from inflammatory carcinoma. Marked pleomorphism, signet-ring cell formation, numerous polymorphonuclear leukocytes. May-Grünwald-Giemsa × 400.

Figure 69.   Aspirate from acute mastitis. Large histiocytes with vacuolated, ill-defined cytoplasm, polymorphonuclear leukocytes. May-Grünwald-Giemsa × 400.

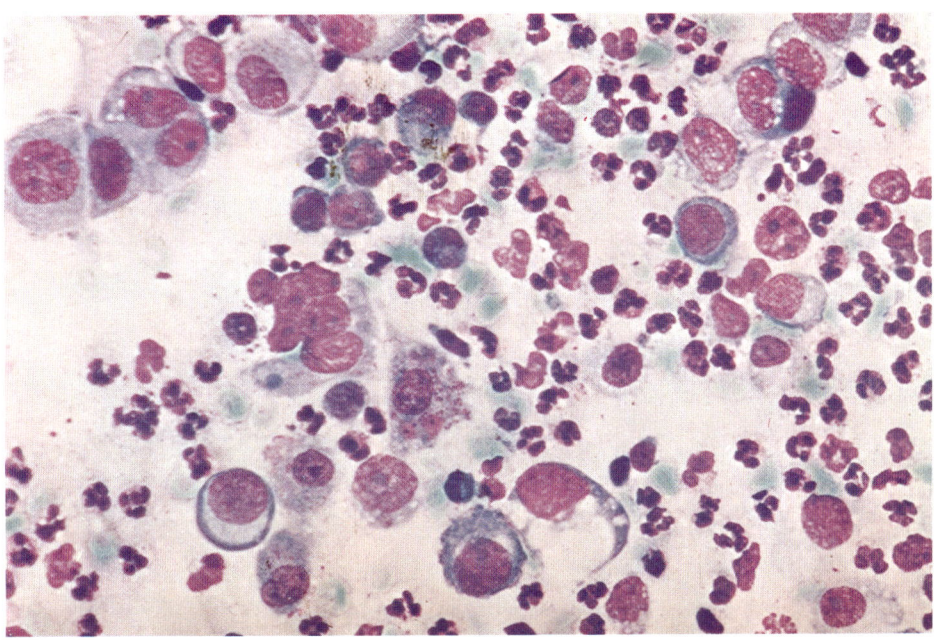

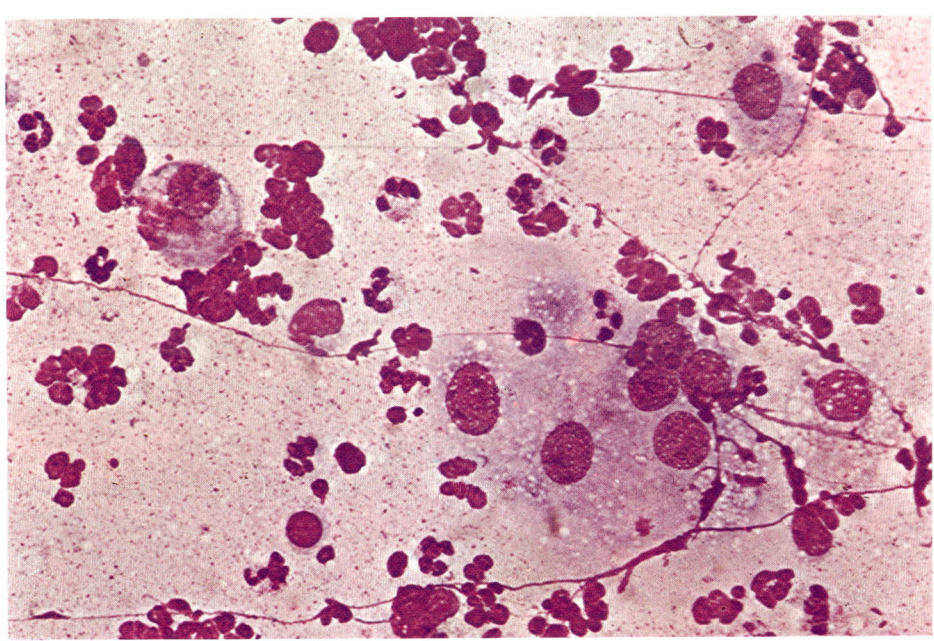

Figure 70. Aspirate from inflammatory carcinoma. Marked pleomorphism, polymorph-onuclear leukocytes. May-Grünwald-Giemsa × 400.

Figure 71. Aspirate from ulcerating carcinoma with inflammatory reaction. Numerous tumor cells and monocytes. May-Grünwald-Giemsa × 400.

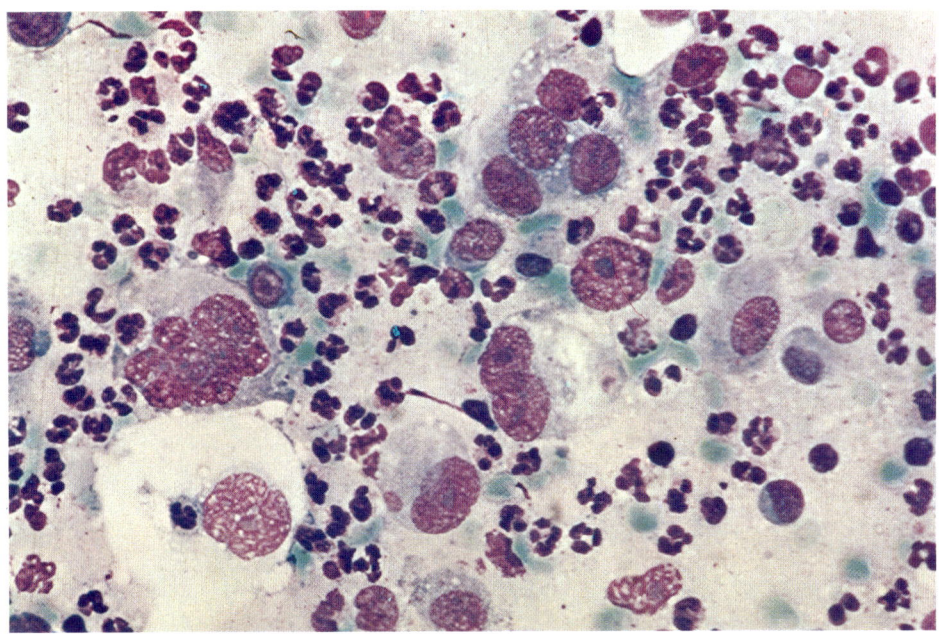

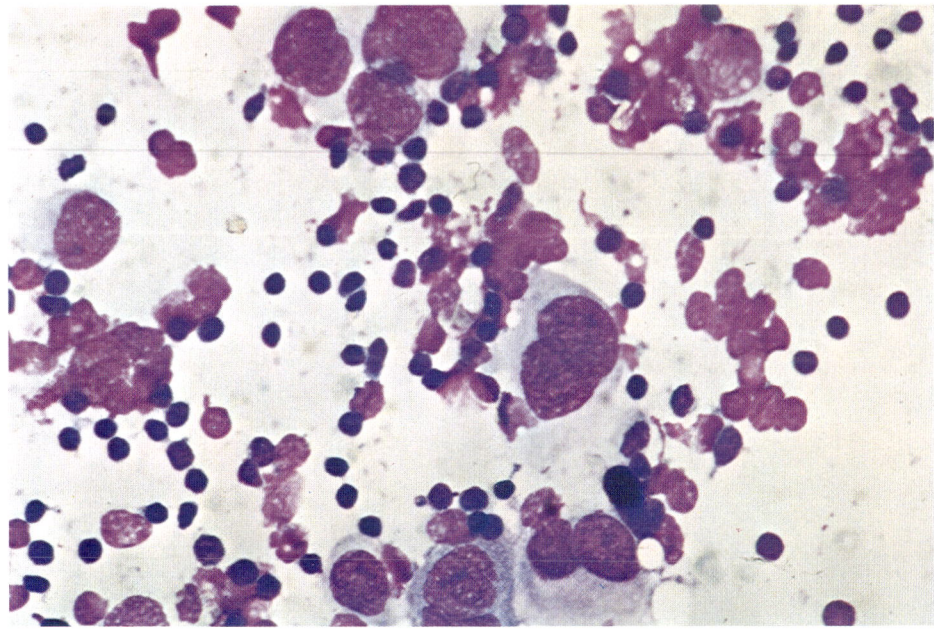

Figure 72. Aspirate from apocrine carcinoma. Malignant cells with moderate pleomorphism, binucleation, and abundant cytoplasm. May-Grünwald-Giemsa × 400.

Figure 73. Aspirate from benign apocrine metaplasia. Compare with Figure 72. May-Grünwald-Giemsa × 400.

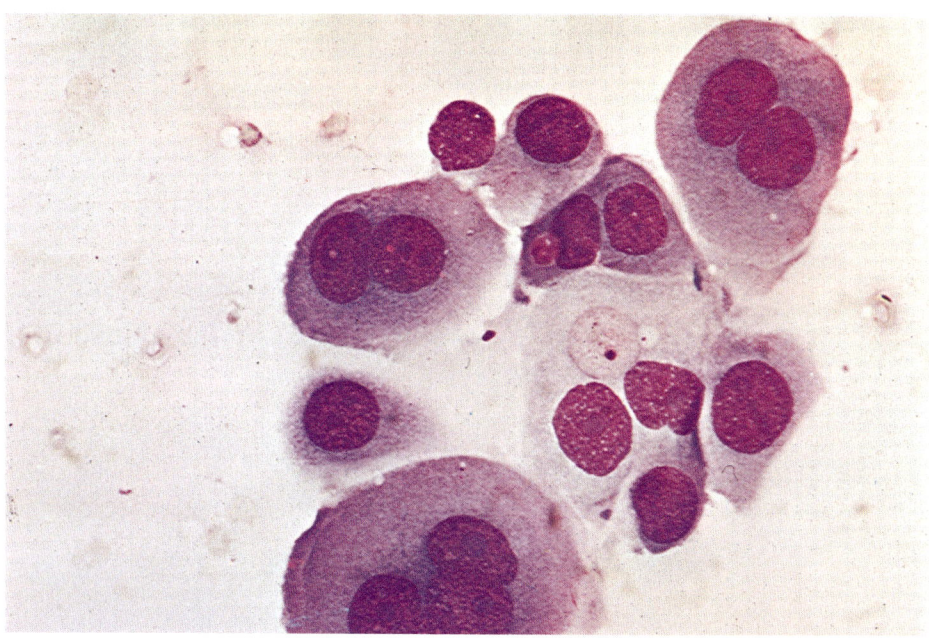

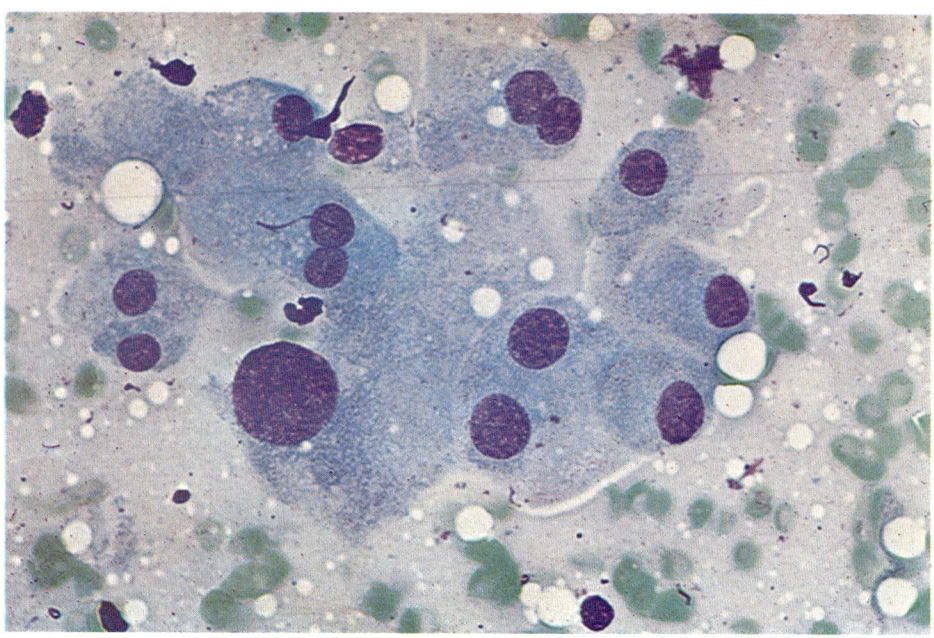

Figure 74. Aspirate from so-called acinic cell carcinoma.[122] Small round tumor cells with scant rim of cytoplasm and prominent nucleoli. May-Grünwald-Giemsa × 400.

Figure 75. Aspirate from a pregnant patient. Enlarged duct cells with prominent nucleoli. Note the similarity to Figure 74. May-Grünwald-Giemsa × 400.

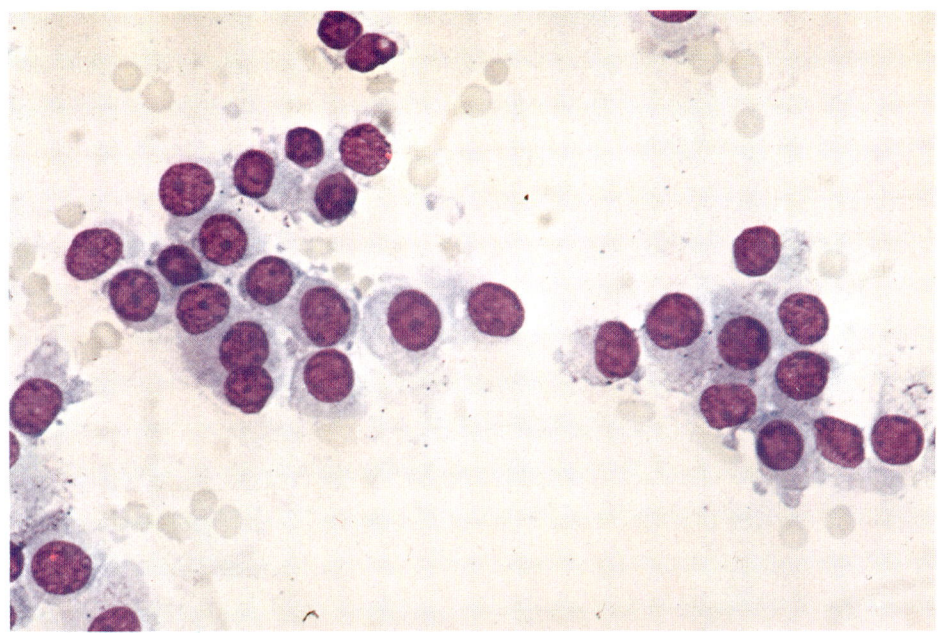

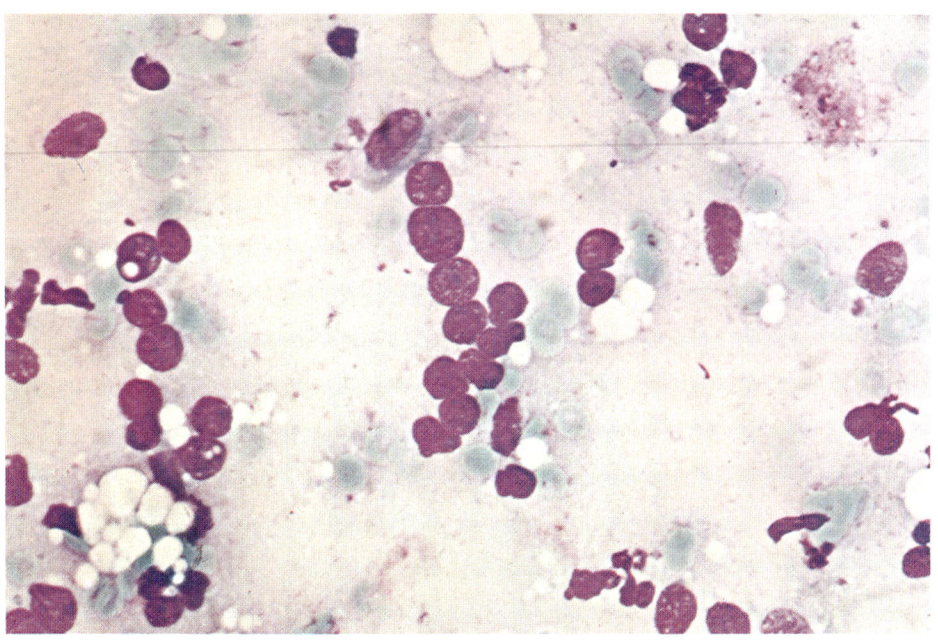

Figure 76. Aspirate from fibrosarcoma of the breast. Several fragments of stromal tissue with oval or spindly malignant nuclei were present in the smear. Papanicolaou × 250.

Figure 77. Tumor cells from fibrosarcoma of the breast. Marked pleomorphism, spindly and oval nuclei, occasional nucleoli. Papanicolaou × 400.

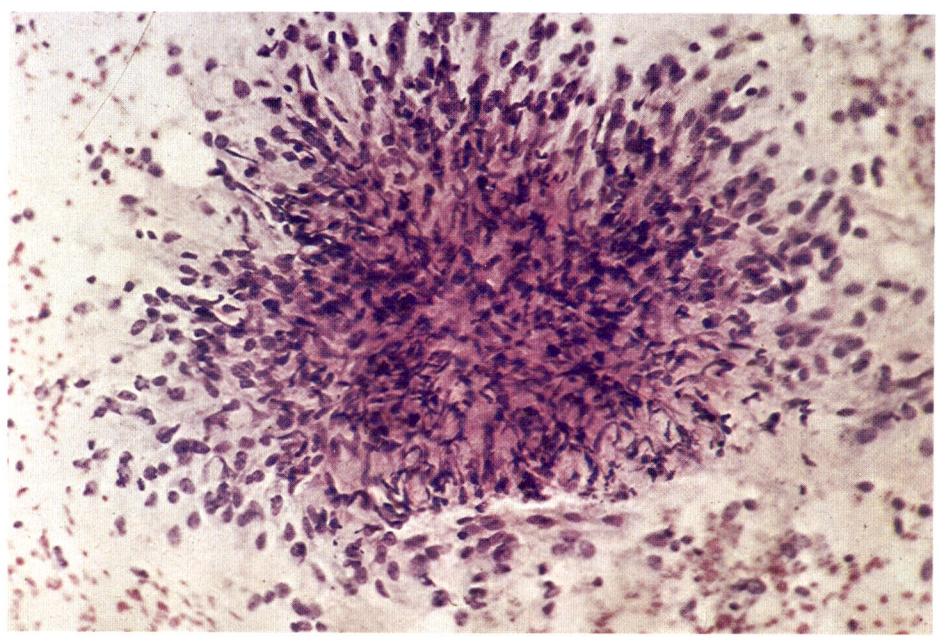

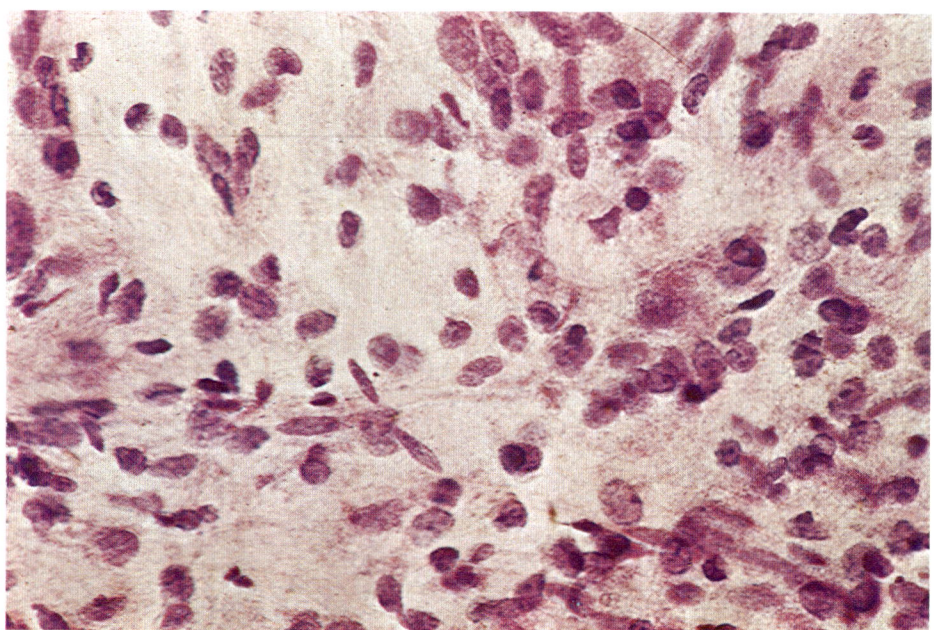

9. Fine needle aspiration of recurrent and metastatic disease

The diagnostic evaluation of metastatic and recurrent carcinoma of the breast by means of needle aspiration is as important and beneficial as the evaluation of primary lesions.[53, 116] Excisional biopsy of locally recurring carcinoma of the breast is often complicated by secondary infection and ulceration as a consequence of previous surgery and radiation. Prevention of poor wound healing and rapid and easy performance of the procedure on an outpatient basis are major advantages of needle aspiration in these circumstances. There is no difference in technique between the aspiration of primary and metastatic lesions. The aspiration of locally recurring carcinoma of the breast may occasionally present some problems because of the small size and firm consistency of the lesion.

Recurrent disease that may be evaluated by needle aspiration includes (1) local recurrence in the scar of the mastectomy, (2) metastatic skin involvement, e.g., on the head, arms, or trunk, and (3) lymph node metastases, e.g., in the axillary or clavicular area.

9.1. Local recurrence

Local recurrence usually manifests itself in the form of solitary or multiple, small flat nodules in the mastectomy scar (Fig. 78). Technically, the aspiration may sometimes be difficult, owing to the extreme flatness of locally recurring lesions. The needle should be inserted tangentially to insure that the tip does not bypass the lesion. Firm immobilization of the nodule with two fingers of one hand is necessary to allow proper performance of the backward and forward movements of the needle, essential for aspiration of cellular material.

Postoperative hematoma, accumulation of serous fluid, foreign body granuloma, and inflammation are differential diagnoses that have to be considered with nodules at this site.

9.2. Metastatic skin involvement

Because of their larger size, metastatic nodules of the skin are not as difficult to aspirate as those that recur locally. In our series, distant skin metastases were found on the skull, scalp, forehead, retroauricular area, neck, back, and upper arms as well as on the thorax and abdomen (Figs. 80 and 81).

9.3. Lymph node metastases

This section deals only with involvement of lymph nodes by metastatic carcinoma of the breast. No attempt is made to cover the entire range of lymph node pathology and its evaluation by fine needle aspiration. Aspiration cytology of lymph nodes is a difficult subject that is practiced only in a small number of institutions.[16, 18, 77, 98, 122] The mixture of various cell types normally present in lymph node aspirates represents the main cause for this limitation. Cells of the reticuloendothelial system are particularly difficult to interpret accurately in needle aspirations.[71, 78]

In the absence of prior histological examination, the evaluation by needle aspiration of malignant disease involving lymph nodes, e.g., lymphoma, Hodgkin's disease, and metastatic carcinoma with unknown primary source, is particularly hampered by some of the disadvantages of the procedure:

1. The assessment of cells removed from the surrounding tissue can be extremely difficult or impossible, as in well-differentiated lymphoma of Hodgkin's disease. The invasive potential can no longer be evaluated.[68, 105]

2. In the presence of lymph node metastases with unknown primary source, needle aspiration is usually unable to give any indication of the primary site.[69, 71, 105]

3. Small metastatic foci in the peripheral sinus may be missed even by multiple aspirations.

However, if the patient presents with a history of histologically confirmed neoplastic disease, needle aspiration can make a reliable contribution to the confirmation of metastatic lymph node involvement. As a matter of fact, the aspiration of lymph nodes in the assessment of metastatic spread of carcinoma of the breast was practiced long before primary lesions of the breast were aspirated.[45, 49] Carcinoma cells differ markedly from the indigenous cell population of lymph nodes. Therefore, the cytological diagnosis of metastatic disease is considered to be rather easy.[16, 71, 105] Characteristic features of metastatic cells are their large size, never reached by any of the cells normally found in lymph nodes, and marked pleomorphism (see page 39).

According to Berg,[6] four factors are mainly responsible for the apparent success of needle aspiration in the diagnosis of metastatic lymph node involvement:

1. Easy accessibility of enlarged lymph nodes for palpation and aspiration.

2. Rich cellularity and poor vascularization of metastatic lymph nodes, resulting in a high yield of tumor cells with little admixture of blood.

3. Easy identification of cellular elements foreign to lymph nodes.

4. In most cases, availability of histological sections of the primary lesion for comparative evaluation.

The accuracy of needle aspiration in the detection of metastatic involvement of lymph nodes by carcinoma of the breast is reported in the literature as being 90 to 95 per cent.[24, 96, 122]

Reasons for false negative results are:

1. Focal involvement of the node, which may be missed by aspiration.
2. Necrosis of the metastatic node.
3. Hyalinization of the lymph node.

Reasons for false positive results are:

1. Erroneous interpretation of large reticuloendothelial cells as tumor cells.
2. Aspiration of non-lymphatic structures, such as branchial cleft cysts and carotid bodies.

Essential for the successful application of needle aspiration in the diagnosis of lymph node metastases is an awareness of its limitations. Thus, excisional biopsy should be recommended in clinically suspicious cases with negative cytological results. It should be kept in mind that needle aspiration is only one procedure in the entire set available for the evaluation of lymph nodes.

Figures 78 through 91 appear in the following section.

Figure 78. Local recurrence in the form of multiple, disseminated nodules measuring 0.4 to 0.6 cm in diameter.

Figure 79. Malignant cells obtained by fine needle aspiration from a local recurrence of carcinoma of the breast. The lesion measured 0.4 cm in diameter. Papanicolaou × 400.

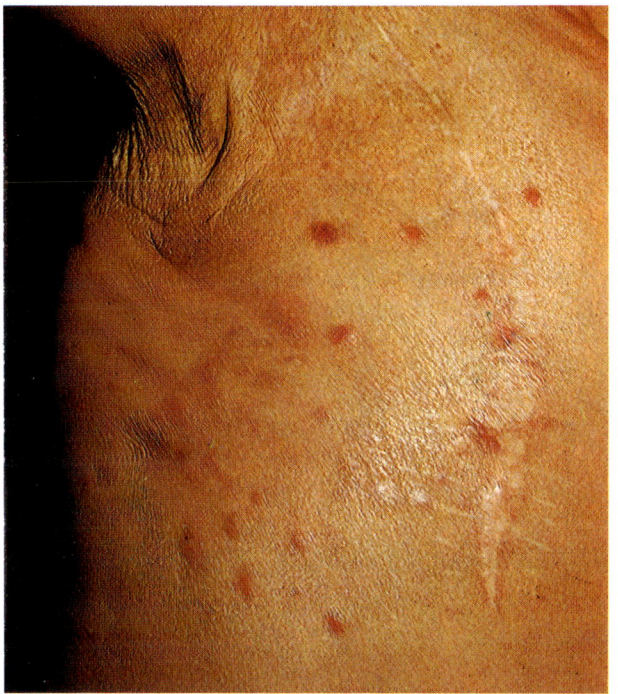

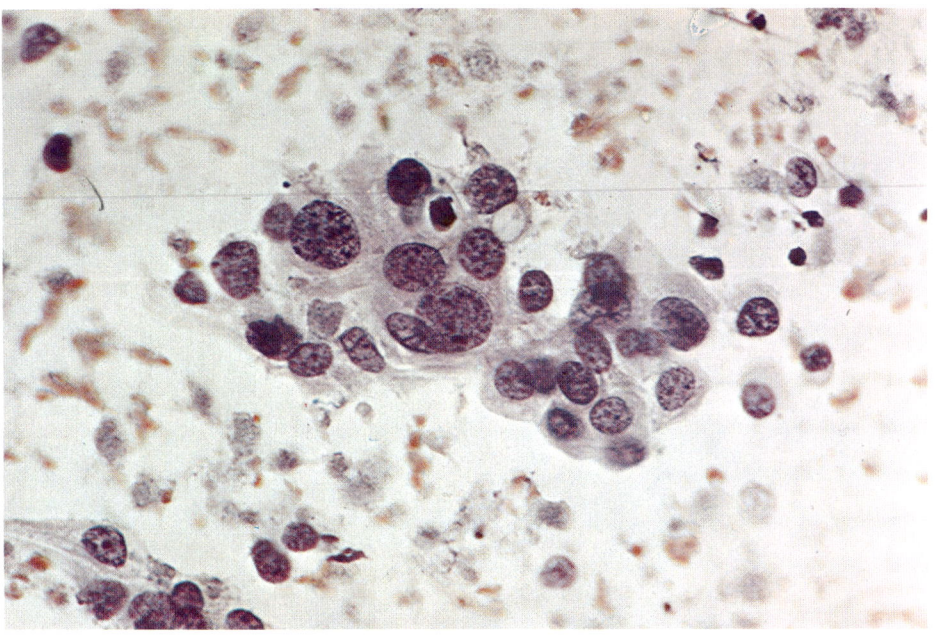

Figure 80.   Tumor cells from a retroauricular metastasis of carcinoma of the breast, that measured 2 cm in diameter. May-Grünwald-Giemsa × 400.

Figure 81.   Tumor cells from a nodular lesion that was attached to a rib. Metastatic carcinoma of the breast. May-Grünwald-Giemsa × 400.

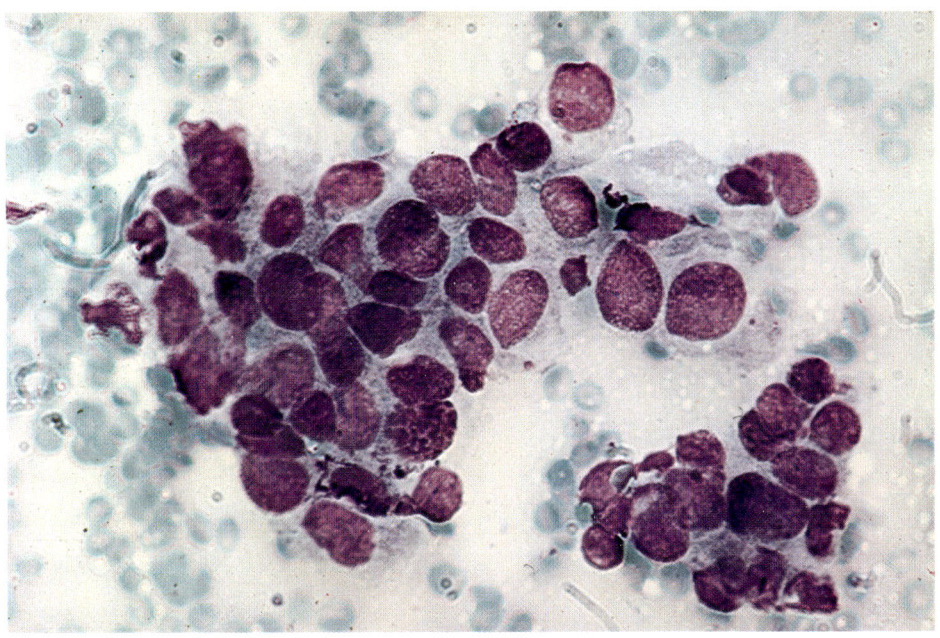

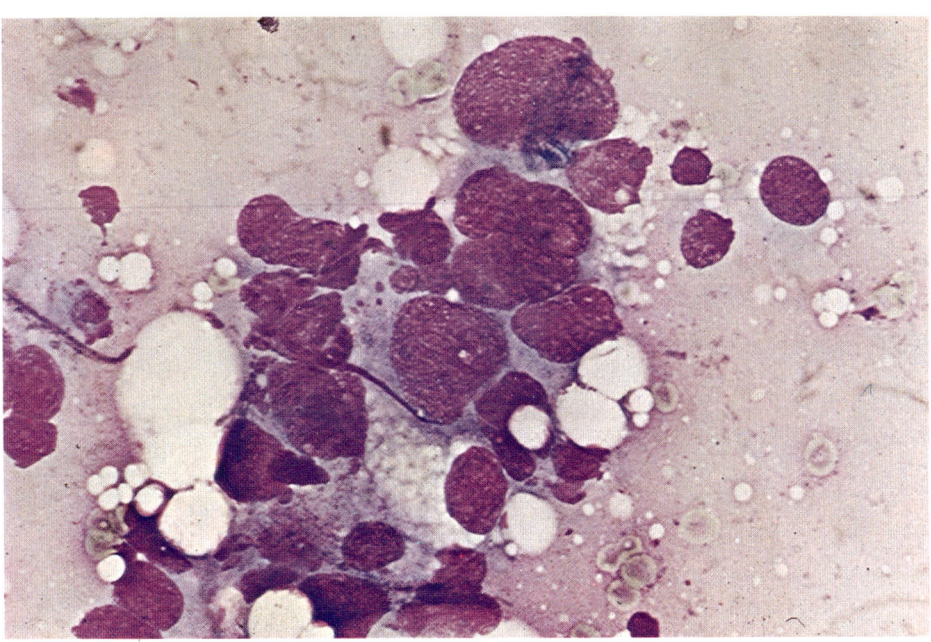

Figure 82. Imprint preparation of a lymph node. Normal cellular elements. May-Grünwald-Giemsa × 400.

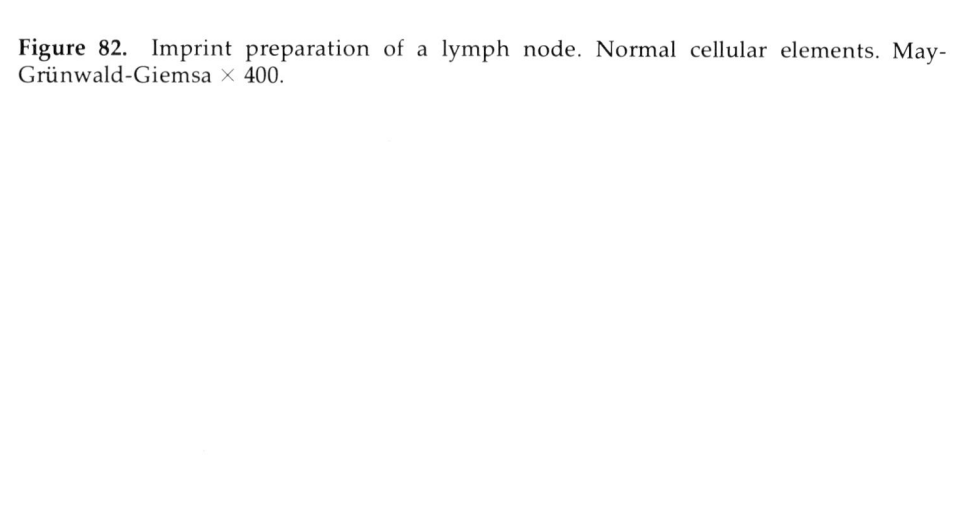

Figure 83. Carcinoma of the breast. Aspiration of a lymph node metastasis. Marked nuclear enlargement and pleomorphism, coarse chromatin, occasional nucleoli. May-Grünwald-Giemsa × 400.

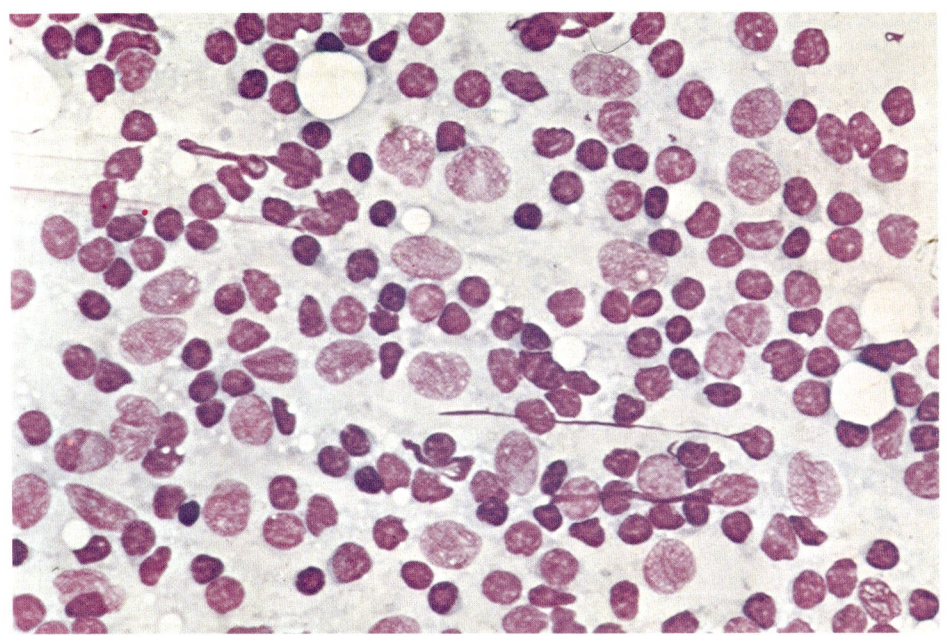

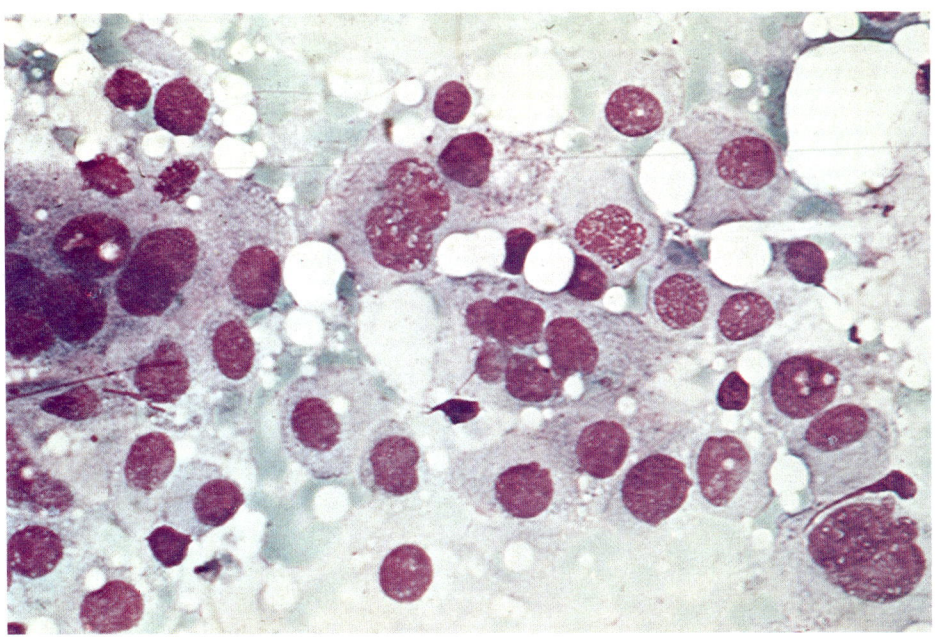

Figure 84.   Aspiration of a lymph node with chronic lymphadenitis. May-Grünwald-Giemsa × 400.

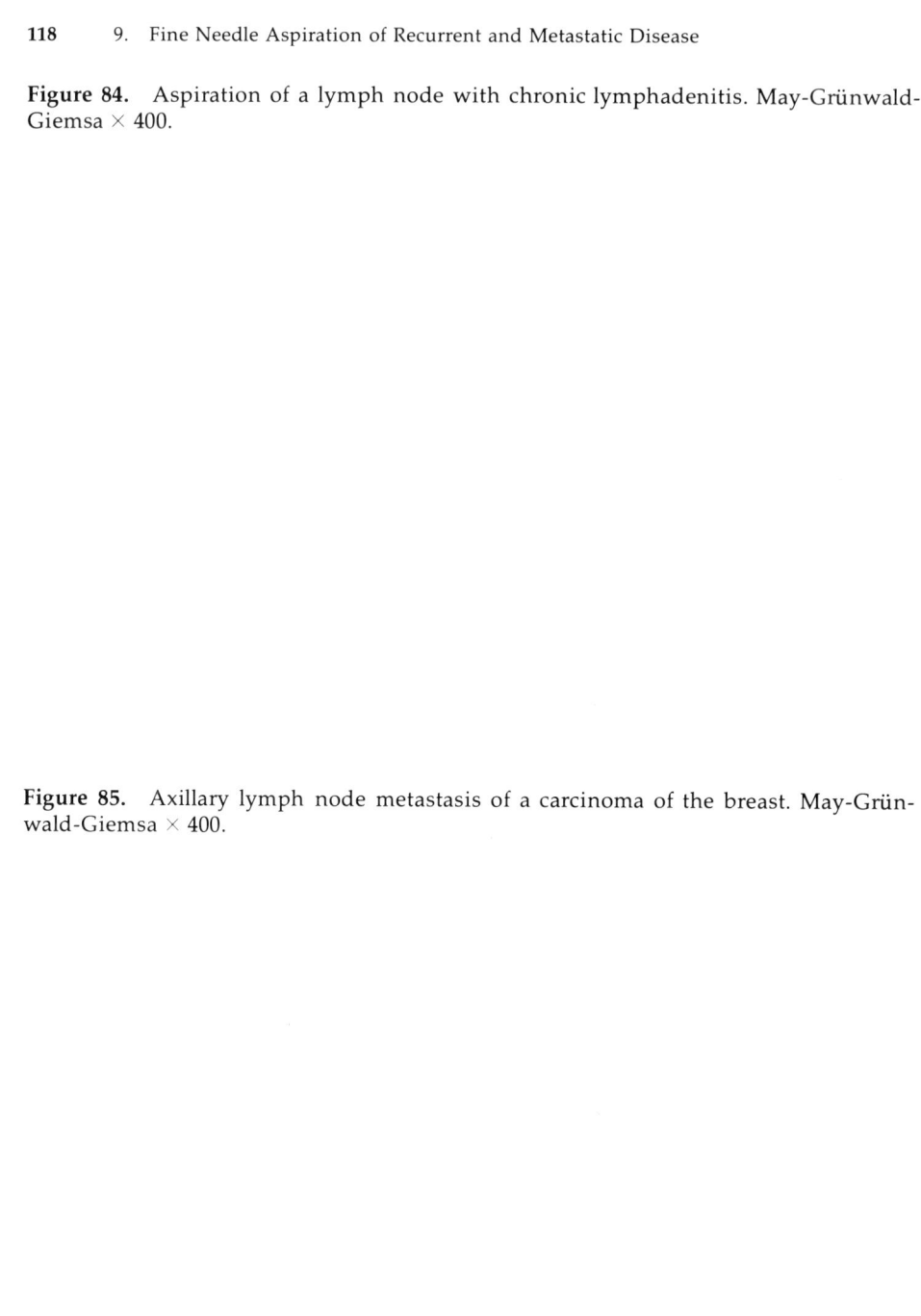

Figure 85.   Axillary lymph node metastasis of a carcinoma of the breast. May-Grünwald-Giemsa × 400.

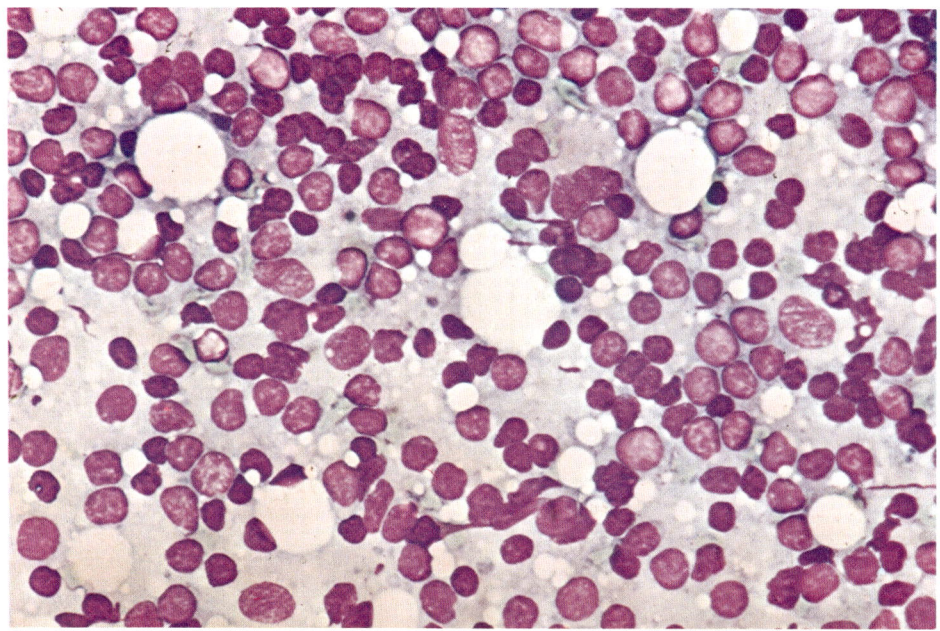

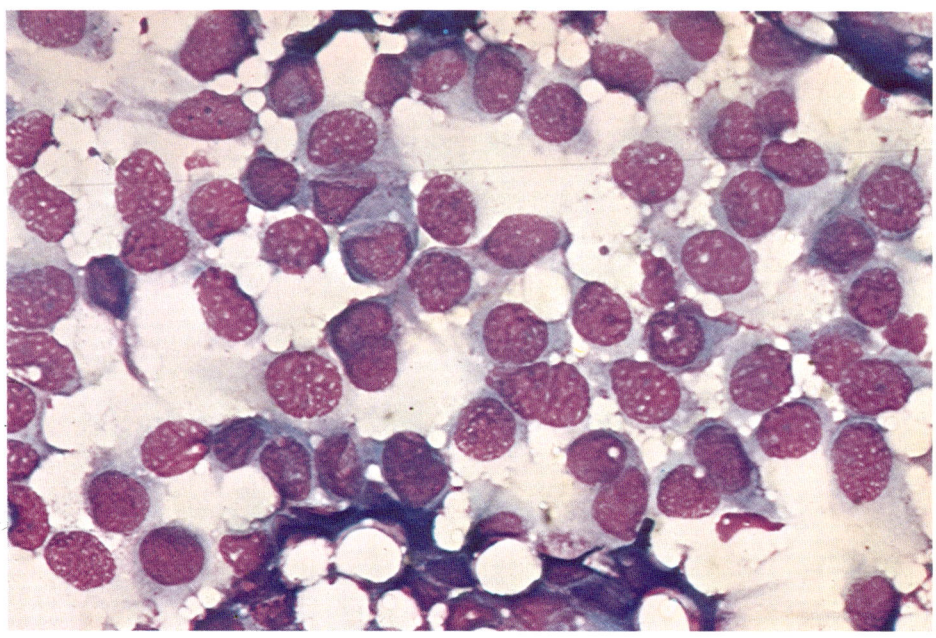

Figure 86.   Aspiration of an axillary lymph node metastasis. Medullary carcinoma of the breast. Papanicolaou × 400.

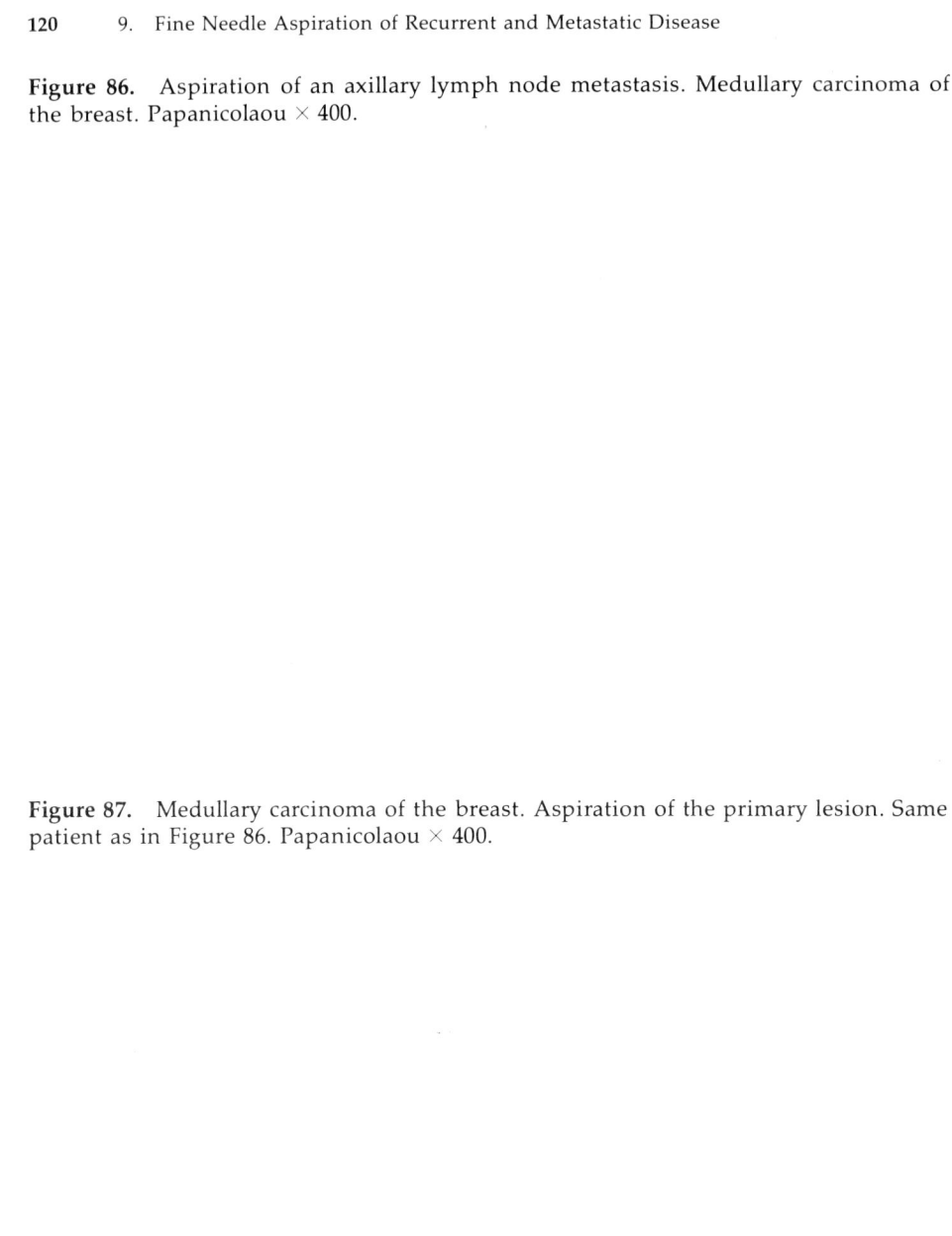

Figure 87.   Medullary carcinoma of the breast. Aspiration of the primary lesion. Same patient as in Figure 86. Papanicolaou × 400.

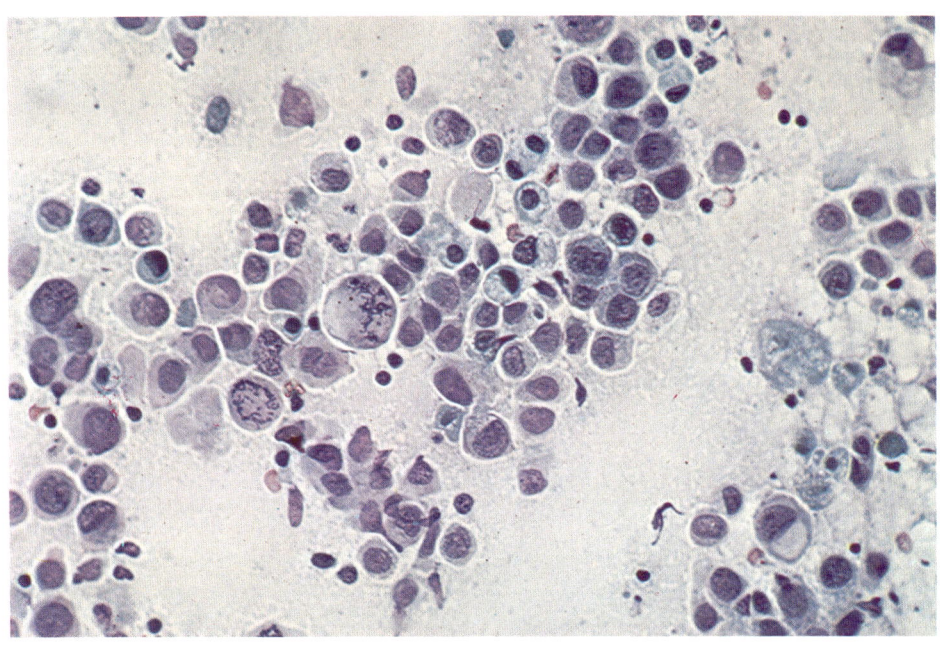

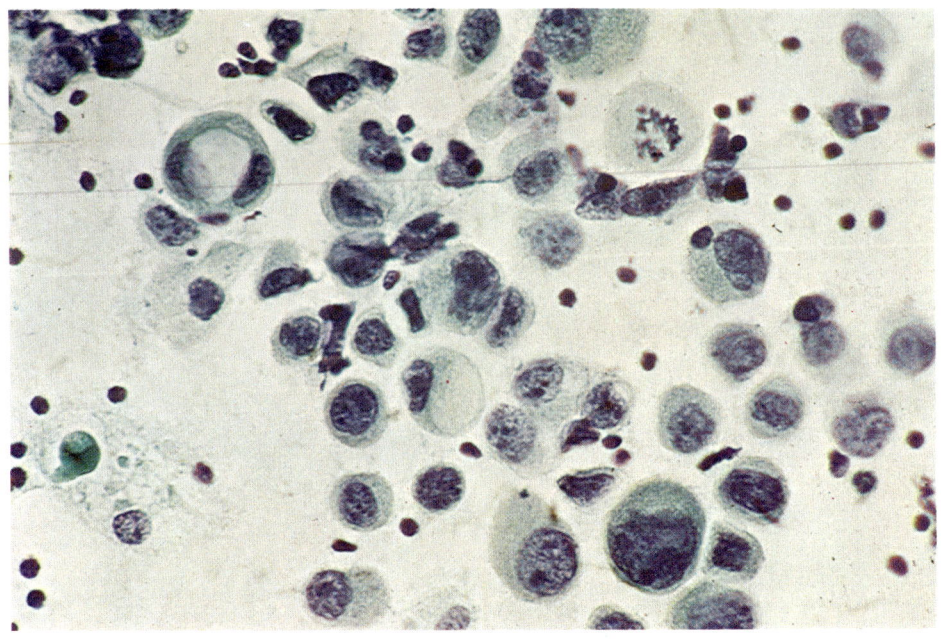

Figure 88.   Aspiration of an inguinal lymph node metastasis. Medullary carcinoma of the breast. Same patient as in Figures 86 and 87. Papanicolaou × 400.

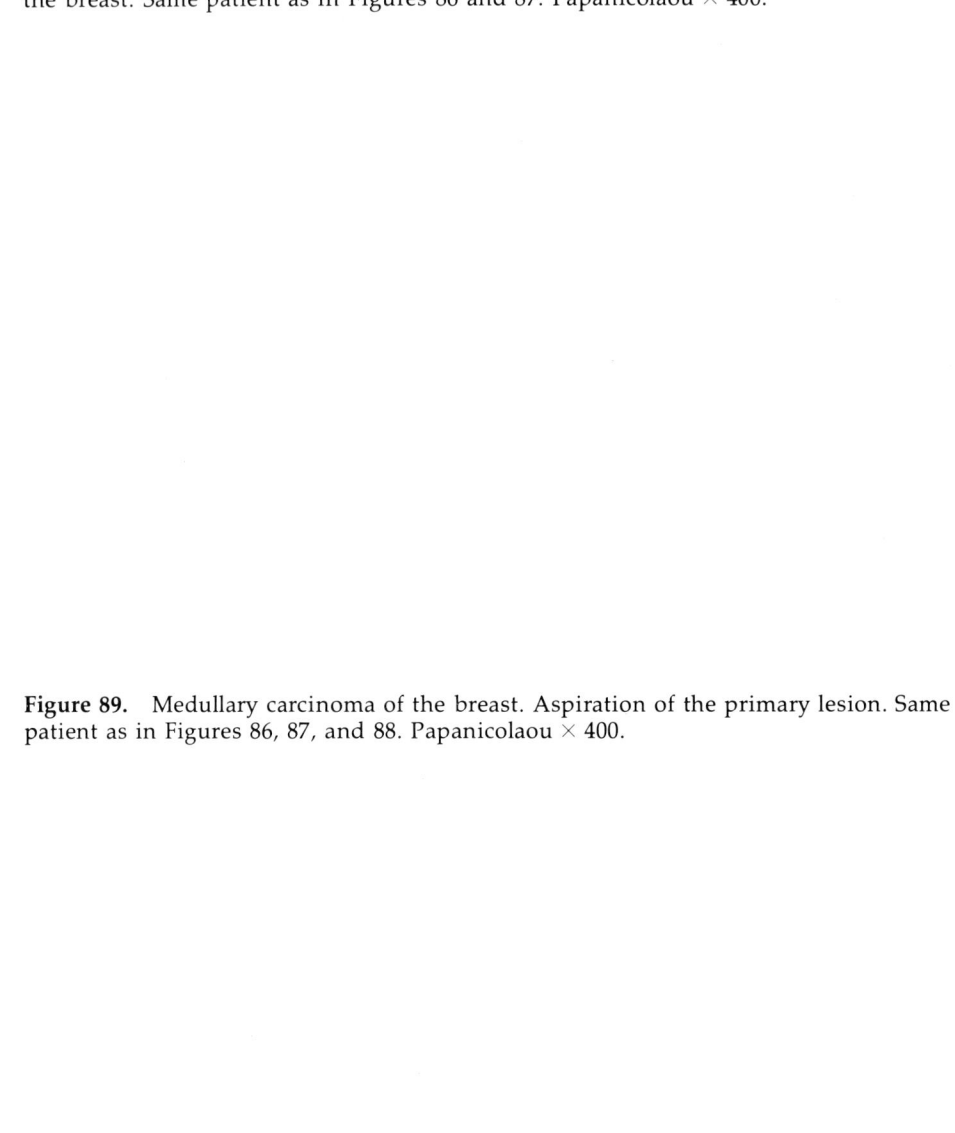

Figure 89.   Medullary carcinoma of the breast. Aspiration of the primary lesion. Same patient as in Figures 86, 87, and 88. Papanicolaou × 400.

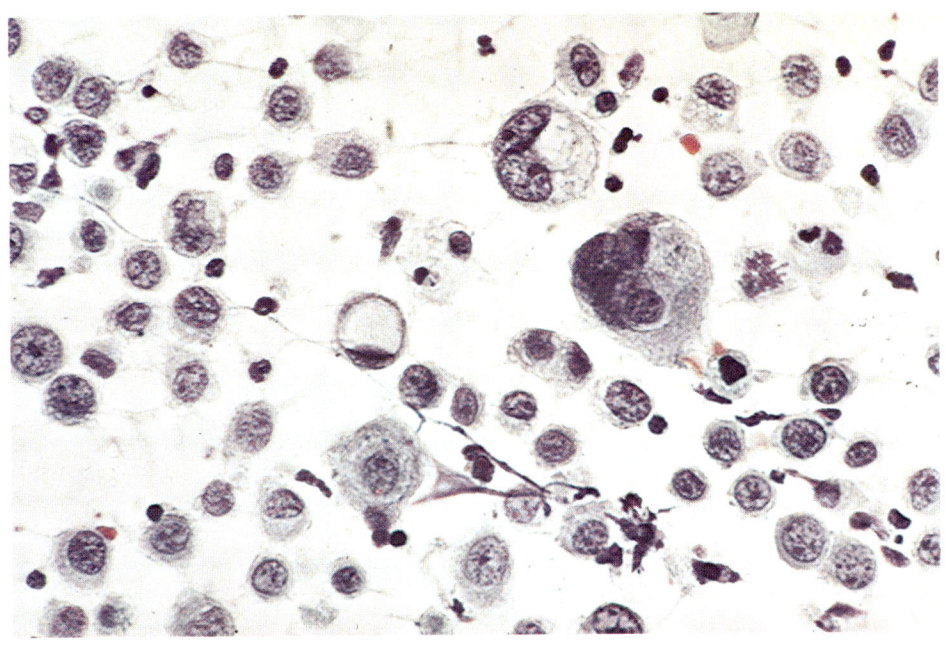

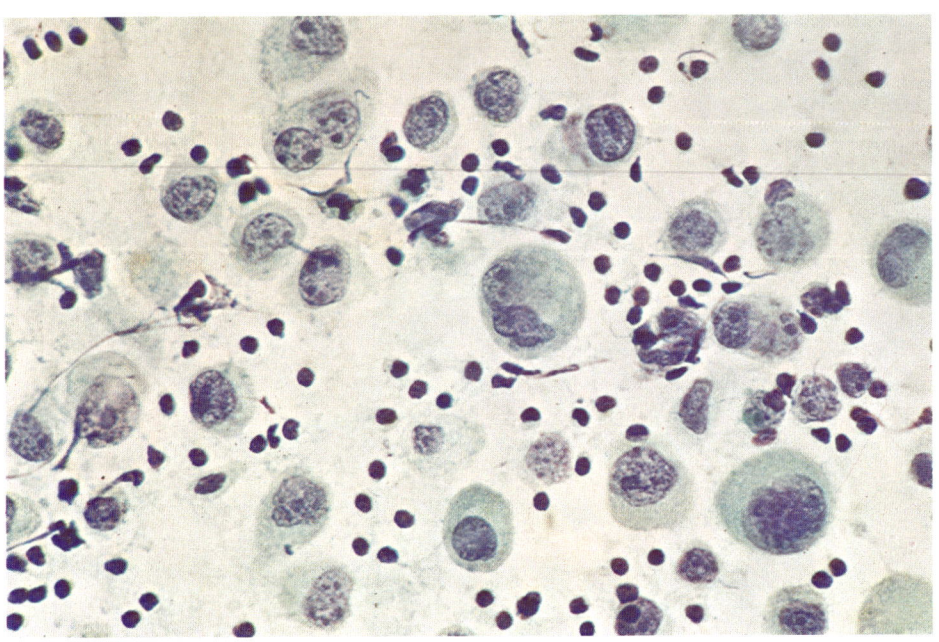

Figure 90.   Aspiration of an axillary lymph node metastasis of apocrine carcinoma. Papanicolaou × 400.

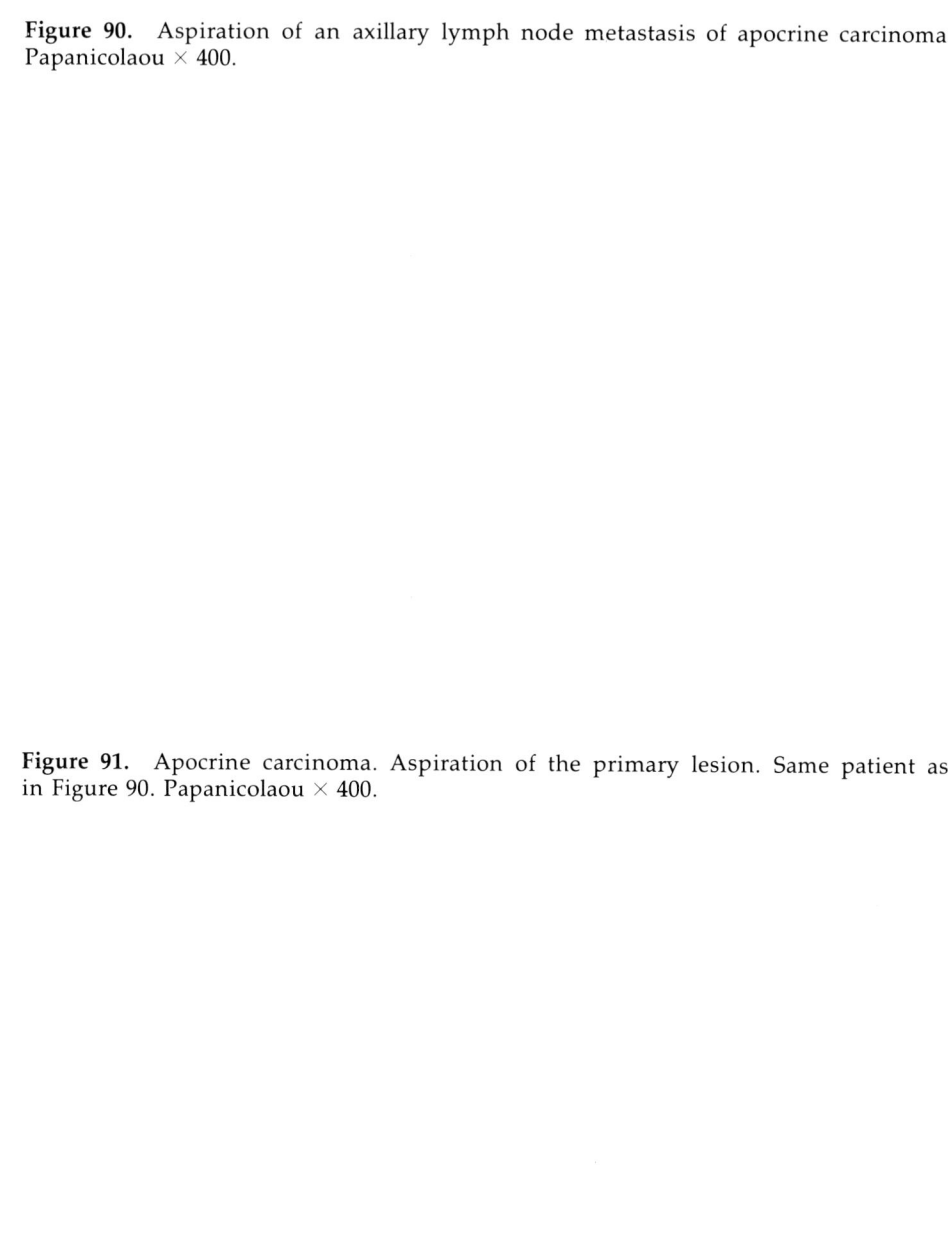

Figure 91.   Apocrine carcinoma. Aspiration of the primary lesion. Same patient as in Figure 90. Papanicolaou × 400.

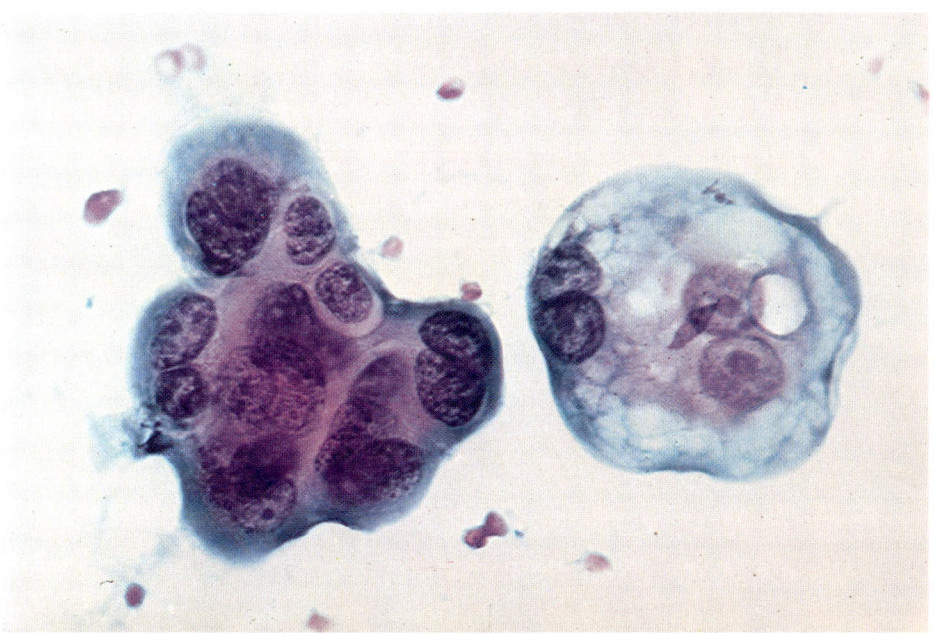

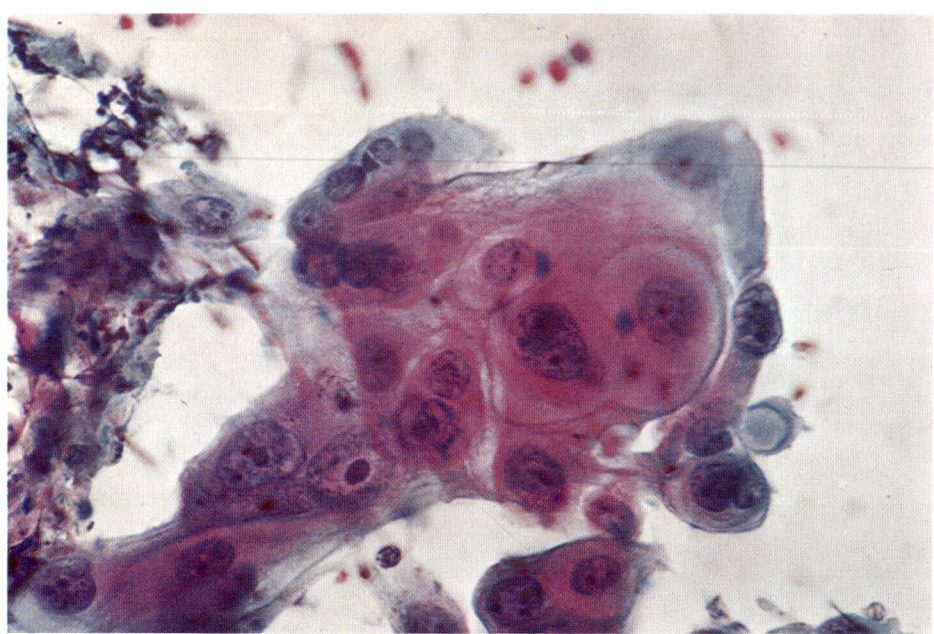

10. Results of fine needle aspiration of the breast

Between May 1971 and May 1976, 2778 lesions of the breast and 230 lymph nodes were aspirated in our department.

TABLE 2. Fine Needle Aspiration of the Breast (May 1971 to May 1976)

Type	Number
Benign tumor	1654
Large cyst	532
Carcinoma	333
Recurrent carcinoma	259
Lymph node	230
Total	3008

Excisional biopsy was necessary in only 212 of 1654 benign tumors of the breast. The remaining lesions were proved to be unremarkable by clinical examination, mammography, thermography, and cytological study. Therefore, surgery was performed only if it was clinically indicated, e.g., in mastitis and abscess formation, foreign body granuloma, and fat necrosis.

For 532 large cysts, needle aspiration was diagnostic and curative at the same time. A second aspiration had to be done in 25 patients, owing to recurrent fluid accumulation after two to six months.

Among 333 malignant tumors of the breast, 14 lesions were detected by cytological examination, after clinical examination, mammography, and thermography all gave negative results. In 29 patients, mammography and thermography both gave negative results, whereas malignant cells were obtained by needle aspiration, confirming the clinical impression of malignancy.

TABLE 3. Comparison of Aspiration Cytology and Histology in Carcinoma of the Breast

Histologically Proven Carcinoma		Results of Aspiration Cytology		
Year	Number of Cases	Positive	Suspicious	False Negative
1971	24	19	3	2
1972	37	31	6	0
1973	57	55	2	0
1974	83	79	3	1
1975	106	99	4	3
Total	307	283	18	6

The rate of diagnostic accuracy of fine needle aspiration in comparison with histological examination is given in Table 3. It should be noted that the majority of false negative or doubtful interpretations date back to the years 1971 and 1972, the first two years after the introduction of this technique in our institution. Three out of six carcinomas that were missed cytologically turned out to be scirrhous carcinoma, which corresponds to reports in the literature.[17, 20, 64, 65, 72] The specimens were poorly cellular as a consequence of the dense desmoplastic reaction characteristic of scirrhous carcinoma that makes the aspiration of sufficient cytological material difficult.

According to Marsan and his associates,[72] reasons for false negative results are:

1. Poor cellularity in aspirates from malignant neoplasias with marked desmoplastic reaction (scirrhous carcinoma).

2. Marked similarity between tumor cells and their benign counterparts as in the small cell type of carcinoma of the breast or carcinoma with monomorphous appearance.

3. Simultaneous occurrence of benign and malignant cells, as in aspirates of poorly delineated lesions.

Small size of the tumor (less than 1 cm in diameter) is cited in the literature as a major reason for false negative results.[64, 90, 113] In our series, this cannot be confirmed. Out of 19 malignant lesions measuring less than 1 cm in diameter and 23 measuring between 1 cm and 1.5 cm, only one carcinoma was read as negative and two as suspicious. Location of the tumor and consistency of the tissue are more important than size, since proper palpation, essential for location and aspiration, is dependent mainly on these two factors.

TABLE 4.   False Positive and False Negative Results in
Fine Needle Aspiration of the Breast

Author	Year	Sample	Carcinoma	False Negative Number	False Negative %	Benign Tumors	False Positive Number	False Positive %
Smith et al.[97]	1959	202	80	19	24	122	3	2.5
Zajicek et al.[121]	1970	2077	1068	106	9.9	1009	1	0.1
Cornillot et al.[21]	1971	2081	1235	62	5	846	15	1.7
Zajdela et al.[116]	1975	2772	1745	63	3.6	1027	3	0.3
Vilaplana et al.[110]	1975	382	121	11	9	261	3	1.1
Furnival et al.[35]	1975	178	51	2	4	126	5	4
Kreuzer*	1976	720	356	41	11.5	504	4	0.8
Geier*	1976	338	179	15	8	159	15	9
Present series	1976	532	333	10	3	199	6	3

*Personal communication.

Nuclear pleomorphism occurring in fibroadenoma and occasionally in fibrocystic disease was mainly responsible for false positive or suspicious interpretations in this series. These alterations may cause considerable difficulty in the cytological evaluation even for an experienced cytopathologist. In one case, markedly pleomorphic histiocytes occurring in fat necrosis were erroneously read as malignant. According to Delarue and Orcel,[22] reasons for false positive results are: pleomorphic histiocytes, inflammatory changes in duct cells, radiation cell changes, and technical artifacts.

The literature reports frequent erroneous interpretations of atypical epithelial cells occurring in fibroadenoma[20, 34, 37, 106] as well as misinterpretation of pleomorphic histiocytes[22, 34, 37, 106] as malignant. Papanicolaou[85] recognized early that atypical histiocytes represent a major pitfall in the cytological examination of the breast.

11. Role of needle aspiration in the diagnosis of breast lesions

11.1. Indications

At this point, having discussed the potential and limitations of the procedure, we shall present a summary of indications for fine needle aspiration of the breast.

A definitive cytological diagnosis can be expected in cases of

1. Solitary lesions, clinically considered to be cysts, benign tumors, or malignant neoplasia. These include large lesions and well delineated small lesions.

2. Clinically obvious cystic lesions (diagnostic-therapeutic aspiration).

3. Solitary or multiple small nodules in the mastectomy scar (local recurrence).

4. Lesions suspect of distant metastasis. These include nodules of the skin and enlarged lymph nodes.

A definitive cytological diagnosis cannot always be expected in the differential diagnosis between inflammatory carcinoma and mastitis, and in the presence of an ill-defined induration, or a multinodular lesion.

11.2. Timing and diagnostic value of needle aspiration

11.2.1. Primary lesion

As a prerequisite, a lesion has to be palpable before needle aspiration can be performed. Moreover, needle aspiration must not be used as a screening procedure for unselected cases but should be regarded as a complementary technique in the entire diagnostic work-up.

Taking a history and inspecting and palpating the lesion are initial steps that serve as guidelines for the application of further diagnostic procedures. Cystic lesions are immediately aspirated if clinically indicated. As already outlined in the section on the management of large cysts (see page 67), no further work-up is necessary if the results of cytological and radiological examinations are negative. The apparent success in the management of 532 cysts in this series supports this view. Non-cystic lesions of unknown nature should first be subject to mammography and thermography, since needle aspiration may occasionally interfere with these procedures, e.g., by producing a hematoma. Subsequently performed needle aspiration will render a definitive diagnosis in 96 to 97 per cent of solid lesions (Fig. 92).

Negative results of needle aspiration

In the presence of a solitary mass, there is no need for excisional biopsy if clinical, radiological, and cytological examination prove to be negative. However, because of the false negative results with needle aspiration in 3 to 4 per cent of cases, the following guidelines were developed in our institution:

1. All palpable lesions are checked at short intervals after a negative work-up. Self-examination may be useful in this situation.

2. Excisional biopsy is indicated, if: (A) the lesion grows larger or changes in consistency, (B) the lesion is initially larger than 2 to 3 cm in diameter, (C) the excision is requested by the patient, or (D) the lesion itself or the conservative management of it creates anxiety in the patient.

The role of fine needle aspiration in the management of diffuse or multiple lesions, e.g., in multinodular mastopathy, is problematical. There is a high margin of error. The same holds true for non-palpable lesions detected by radiological techniques. In both cases, needle aspiration does not represent a reliable technique.

Positive results of needle aspiration

In .the case of an unequivocal cytological diagnosis of malignancy, the patient should be scheduled for immediate surgery. Even if all other examinations prove to be negative, the probability of the presence of a malignant tumor is in the range of 96 to 97 per cent. At surgery, an excisional biopsy should be done. Examination of a frozen section is recommended for confirmation of the diagnosis and for legal reasons. It is the opinion in our laboratory that a frozen section procedure could be omitted in a large number of cases that show a positive result from the needle aspirate (85 per cent).

11.2.2.   Recurrent disease

Needle aspiration plays an important role in the diagnostic evaluation of recurrent and metastatic disease and is indicated for

1. Solitary or multiple lesions in the mastectomy scar.

2. Axillary masses, which may represent local recurrence or metastatic lymph nodes.

3. Suspected lymph node metastases, e.g., in the neck or clavicular area.

4. Suspected distant metastases, e.g., in skin, remaining breast, or musculature.

In areas of previous surgery or radiation, on which biopsies are only reluctantly performed, needle aspiration represents a reliable alternative technique. However, in the evaluation of non-palpable, ill-defined metastatic spread, e.g., lymphangiosis carcinomatosa or early spray-like metastases in the mastectomy scar, needle aspiration is of limited value and excisional biopsy is preferred.

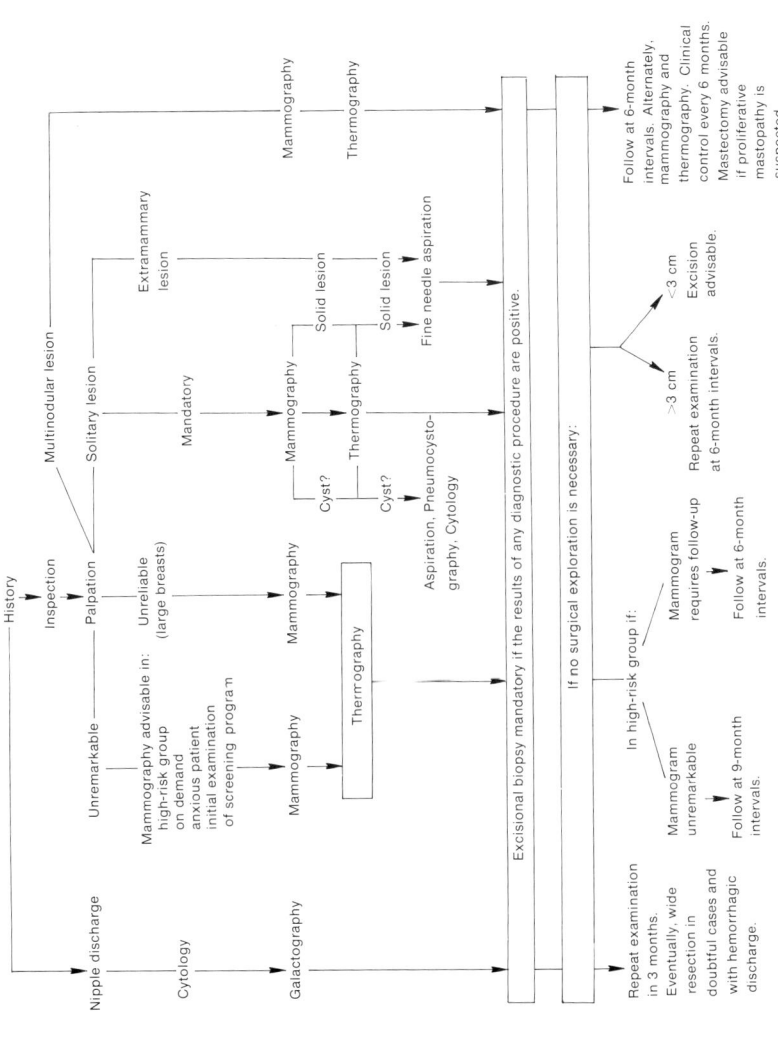

Figure 92.   Role of fine needle aspiration in the diagnostic evaluation of breast lesions. (*From* Schmidt-Matthiesen, H., and Glätzner, H. *In:* Der Frauenarzt 16:192, 1975)

References

1. Ashton, P. R., Hollingsworth, A. S., and Johnston, W. W.: The cytopathology of metastatic breast cancer. Acta Cytol., 19:1, 1975.

2. Barnes, W. C.: Management of cystic disease of the breast. Am. J. Surg., 129:324, 1975.

3. Barth, V., Müller, R., Deininger, H. K., and Wöllgens, P.: Technische Fehlermöglichkeiten bei der erweiterten Mammadiagnostik. Dtsch. Med. Wochenschr., 98:1724, 1973.

4. Barth, V., Müller, R., Deininger, H. K., and Wöllgens, P.: Klinik, Mammographie, Zytologie, Stanzbiopsie und Plattenthermographie in der erweiterten Mammadiagnostik. Dtsch. Med. Wochenschr., 99:175, 1974.

5. Bässler, R.: Zur Definition und Dignität des Carcinoma in situ der Brustdrüse. Österr. Z. Onkol., 2:125, 1975.

6. Berg, J. W.: The Aspiration Biopsy Smear. In: Koss, L. G.: Diagnostic Cytology and its Histopathologic Bases. 1st ed. Philadelphia, J. B. Lippincott, 1961, p. 311.

7. Berg, J. W., and Robbins, G. F.: A late look at the safety of aspiration biopsy. Cancer, 15:826, 1962.

8. Bethel, E. P. A.: La Mammacytologie. Paris, D. P. Taib, 1958.

9. Bibbo, M., and Zuspan, F. P.: Fine-needle aspiration of the breast in an obstetrics and gynecology hospital. Am. J. Obstet. Gynecol., 122:525, 1975.

10. Bonneau, H., Sommer, D., and De The, G.: Etude critique de la cytologie tumorale par ponction à l'aiguille fine. Presse Med., 68:909, 1960.

11. Boquoi, E., and Kreuzer, G.: Punktionszytologie der Mamma. Dtsch. Ärztebl., 45: 3227, 1974.

12. Boschann, H. W.: Gynäkologische Zytodiagnostik für Klinik und Praxis 2. Auflage Berlin, Walter de Gruyter, 1973.

13. Bothmann, G., Rummel, H., and Kubli, F.: Punktionzytologie bei Erkrankungen der weiblichen Brustdrüse. Diagnostik, 7:791, 1974.

14. Buchwald, W.: Die Differentialdiagnose der voroperierten Brustdrüse in der Mammographie: Narbe-Narbenkarzinom. Fortschr. Röntgenstr., 105:857, 1966.

15. Budd, J. W.: Evaluation of needle aspiration technic in breast lesions. Radiology, 52:502, 1949.

16. Cardozo, P. L.: The cytologic diagnosis of lymph node punctures. Acta Cytol., 8:194, 1964.

17. Castelain, G., and Castelain, C.: Importance et valeur relative des différents caractères cytologiques de malignité cellulaire. Presse Med., 63:764, 1955.

18. Castelain, G., and Castelain, C.: Possibilités et limites du cyto-diagnostic extemporané. Presse Med., 61:1020, 1953.

19. Chu, E. W., and Hoye, R. C.: The clinician and the cytopathologist evaluate fine needle aspiration cytology. Acta Cytol., 17:413, 1973.

20. Cornillot, M., and Verhaeghe, M.: Données cytologiques dans les ponctions de tumeurs du sein. Pathol. Biol., 7:793, 1959.

21. Cornillot, M., Verhaeghe, M., Cappelaere, P., and Clay, A.: Place de la cytologie par ponction dans le diagnostic des tumeurs du sein. Lille Med., 16:1027, 1971.

22. Delarue, J., and Orcel, L.: Etudes des resultats obtenus par les techniques cytologiques et leurs principales causes d'erreurs. Sem. Hop. Paris, *28*:2035, 1952.

23. Eisen, M. J., and Taft, R. H.: Cytological diagnosis of mammary cancer associated with incipient Paget's disease of the nipple. Cancer, *4*:150, 1951.

24. Engzell, U., Jakobsson, P. A., Sigurdson, A., and Zajicek, J.: Aspiration biopsy of metastatic carcinoma in lymph nodes of the neck. Acta Otolaryngol., *72*:138, 1971.

25. Engzell, U., Esposti, P. L., Rubio, C., Sigurdson, A., and Zajicek, J.: Investigation on tumour spread in connection with aspiration biopsy. Acta Radiol., *10*:385, 1971.

26. Esposti, P. L., Franzen, S., and Zajicek, J.: The Aspiration Biopsy Smear. *In*: Koss, L. G., Diagnostic Cytology and its Histopathologic Bases, 2nd ed. Philadelphia, J. B. Lippincott, 1968, p. 565.

27. Evers, R., and Fischedick, O.: Die Punktionsbiopsie des Mammakarzinoms. Fortschr. Röntgenstr., *118*:466, 1973.

28. Fassin, Y., and Jeanmart, L.: La ponction des kystes du sein. Rapport de la pneumonmastographie. J. Belge Radiol., *55*:39, 1972.

29. Fiebelkorn, H. J.: Ist die Diagnose des Mammakarzinoms zytologisch möglich? Strahlentherapie, *95*:587, 1954.

30. Finsterer, H., Prechtel, K., and Doletschek, C.: Vergleichende zyto- und histomorphologische Untersuchungen umschriebener krankhafter Brustdrüsenveränderungen. Geburtshilfe Frauenheilkd., *33*:173, 1973.

31. Fischer, E., and Braun, J.: Seltene benigne zystische Veränderungen der weiblichen Brust. Fortschr. Röntgenstr., *118*:207, 1973.

32. Fogher, L.: Etude de la cytologie des tumeurs du sein chez la femme. Memoire d'assistant étranger. Paris, *104*:29, 1954.

33. Fournier, D.: Die Frühdiagnose des Mammacarcinoms. Hessisches Ärzteblatt, *8*:607, 1975.

34. Franzen, S., and Zajicek, J.: Aspiration biopsy in diagnosis of palpable lesions of the breast. Acta Radiol., *7*:241, 1968.

35. Furnival, C. M., Hughes, H. E., Hocking, M. A., Reid, M. M. W., and Blumgart, L. H.: Aspiration cytology in breast cancer, its relevance to diagnosis. Lancet, *2*:446, 1975.

36. Gatchell, F. G., Dockerty, M. B., and Clagett, O. T.: Intracystic carcinoma of the breast. Surg. Gynecol. Obstet., *106*:347, 1958.

37. Geier, G., and Schuhmann, R.: Die Punktionszytologie im Rahmen der Diagnostik tastbarer Mammaveränderungen. Geburtshilfe Frauenheilkd., *34*:294, 1974.

38. Geier, G., Schuhmann, R., and Kraus, H.: Mammapunktionszytologie. Beitr. Pathol., *156*:223, 1975.

39. Gershon-Cohen, J.: Atlas of Mammography. Berlin, Springer, 1970.

40. Gibson, A., and Smith, G.: Aspiration biopsy of breast tumours. Br. J. Surg., *45*:236, 1957.

41. Glassman, J. A.: Aspiration biopsy for detection of carcinoma of the breast: A critique. J. Int. Coll. Surg., *36*:195, 1961.

42. Godwin, J. T.: Aspiration biopsy: Technique and application. Ann. N.Y. Acad. Sci., *63*:1348, 1956.

43. Godwin, J. T.: Cytologic diagnosis of aspiration biopsies of solid or cystic tumours. Acta Cytol., *8*:206, 1964.

44. Goode, J. V., McNeill, J. P., and Gordon, C. E.: Routine aspiration of discrete breast cysts. Arch. Surg., *70*:686, 1955.

45. Greig, E. D. W., and Gray, A. C. H.: Note on the lymphatic glands in sleeping sickness. Lancet, *1*:1570, 1904.

46. Gros, C.: Radioklinische Diagnose des Mammacarcinoms. Röntgen-Bl., *13*:373, 1960.

47. Gros, C., Gautherie, M., Bourjat, P., and Girardie, J.: Necessity of complementary investigation methods for early diagnosis of breast cancer. Arch. Geschwulsforsch., *39*:304, 1972.

48. Grundmann, E.: Allgemeine Cytologie. Stuttgart, Thieme, 1964.

49. Guthrie, C. G.: Gland puncture as a diagnostic measure. Bull. Johns Hopk. Hosp., *32*:266, 1921.

50. Haage, H., and Fischedick, O.: Die Solitärzyste der weiblichen Brust im Röntgenbild. Fortschr. Röntgenstr., *100*:639, 1964.

51. Haagensen, C. D.: Diseases of the Breast. 2nd ed. Philadelphia, W. B. Saunders, 1974.

52. von Haam, E.: A comparative study of the accuracy of cancer cell detection by cytological methods. Acta Cytol., 6:508, 1962.

53. Hajdu, S. I., and Melamed, M. R.: The diagnostic value of aspiration smears. Am. J. Clin. Pathol., 59:350, 1973.

54. Hébert, G., and Ouimet-Oliva, D.: Diagnosis and management of breast cysts. Am. J. Roentgenol. Radium Ther. Nucl. Med., 115:801, 1972.

55. Hennig, K., Johansson, H., Rimsten, A., and Stenkvist, B.: X-ray and fine-needle biopsy in diagnosis of non-palpable breast lesions. Acta Cytol., 19:7, 1975.

56. Hermann, J. B.: Mammary cancer subsequent to aspiration of cysts in the breast. Ann. Surg., 173:40, 1971.

57. Hoeffken, W., and Hintzen, C.: Die Diagnostik der Mammazysten durch Mammographie und Pneumocystographie. Fortschr. Röntgenstr., 112:9, 1970.

58. Hofmann, W. D., and Kern, G.: Die mikroskopische Untersuchung von Brustdrüsensekreten. Geburtshilfe Frauenheilkd., 30: 525, 1970.

59. Johnston, J. H.: Aspiration as diagnostic and therapeutic procedure in cystic disease of the breast. Ann. Surg., 139:635, 1954.

60. Kern, W. H., Dermer, G. B., and Tiemann, R. M.: Comparative morphology of histiocytes from various organ systems. Acta Cytol., 14:205, 1970.

61. Kern, W. H., and Dermer, G. B.: The cytopathology of hyperplastic and neoplastic mammary duct epithelium. Acta Cytol., 16:120, 1972.

62. Kline, T. S., and Lash, S.: Nipple secretion in pregnancy. Am. J. Clin. Pathol., 37:626, 1962.

63. Koss, L. G.: Diagnostic Cytology and its Histopathologic Bases. 2nd ed. Philadelphia, J. B. Lippincott, 1968.

64. Kreuzer, G., and Zajicek, J.: Cytologic diagnosis of mammary tumours from aspiration biopsy smears. III. Studies on 200 carcinomas with false negative or doubtful cytologic reports. Acta Cytol., 16:249, 1972.

65. Kreuzer, G., Boquoi, E., and Meyer, R. D.: Diagnostik gut- und bösartiger Mammatumoren. Dtsch. Med. Wochenschr., 98:691, 1973.

66. Kreuzer, G., and Boquoi, E.: Die Tripeldiagnostik gut- und bösartiger Mammatumoren. (Klinik, Mammographie, Zytologie.) Geburtshilfe Frauenheilkd., 34:279, 1974.

67. Leborgne, R.: Diagnosis of tumours of the breast by simple roentgenography. Am. J. Roentgenol. Radium Ther. Nucl. Med., 65: 1, 1951.

68. Leiber, B.: Der menschliche Lymphknoten. Munich, Urban u. Schwarzenberg, 1961.

69. Lennert, K.: Pathologie der Halslymphknoten. Arch. Ohr. Nas. Kehlkopfheilk., 182:1, 1963.

70. Linsk, J., Kreuzer, G., and Zajicek, J.: Cytologic diagnosis of mammary tumours from aspiration biopsy smears. II. Studies on 210 fibroadenomas and 210 cases of benign dysplasia. Acta Cytol., 16:130, 1972.

71. Lorenz, W.: Über die Bedeutung der Lymphknotenpunktion für die Strahlenheilkunde. Strahlentherapie, 79:435, 1949.

72. Marsan, C., and Bertini, B.: La place des methodes cytologiques dans le diagnostic des tumeurs du sein. Pathol. Biol., 8:343, 1960.

73. Martin, H. E., and Ellis, E. B.: Biopsy by needle puncture and aspiration. Ann. Surg., 92:169, 1930.

74. Martin, H. E., and Ellis, E. B.: Aspiration biopsy. Surg. Gynecol. Obstet., 59:578, 1934.

75. Martin, H. E., and Stewart, F. W.: The advantages and limitations of aspiration biopsy. Am. J. Roentgenol. Radium Ther. Nucl. Med., 35:245, 1936.

76. Menges, V., Engeler, V., and Stadelmann, R.: Der diagnostische und therapeutische Wert der Pneumocystographie der Brust. Geburtshilfe Frauenheilkd., 34:909, 1974.

77. Morrison, M., Samwick, A. A., Rubinstein, J., Stich, M., and Loewe, L.: Lymph node aspiration. Am. J. Clin. Pathol., 22:255, 1952.

78. Moeschlin, S.: Beitrag zur Morphologie der reticuloendothelialen Zellen des intravitalen Lymphknotenpunktats. Folia Haematol., 65:181, 1941.

79. Moulonguet, P.: Diagnostic des tumeurs du sein. Presse Med., 59:769, 1951.

80. Mouriquand, J., and Dargent, M.: L'em-

preinte mammaire: Etude cytopathologique. Bull. Cancer, 44:449, 1957.

81. Mühlberger, G., and Lauth, G.: Feinnadelpunktion von 184 Mamma-Zysten. Munch. Med. Wochenschr., 117:947, 1975.

82. Mühlow, A.: A device for precision needle biopsy of the breast at mammography. Am. J. Roentgenol. Radium Ther. Nucl. Med., 121:843, 1974.

83. Murad, T. M., and Snyder, M. E.: The diagnosis of breast lesions from cytologic material. Acta Cytol., 17:418, 1973.

84. Murad, T. M., and von Haam, E.: Ultrastructure of myoepithelial cells in human mammary gland tumours. Cancer, 21:1137, 1968.

85. Papanicolaou, G. N., Holmquist, D. G., Bader, G. M., and Falk, E. A.: Exfoliative cytology of the human mammary gland and its value in the diagnosis of cancer and other diseases of the breast. Cancer, 11:377, 1958.

86. Rehm, A., Amirfallah, M., and Fischedick, O.: Rundherde der Mamma. Radiologe, 10:149, 1970.

87. Robbins, G. F., Brothers, J. H., III, Eberhardt, W. F., and Quan, S.: Is aspiration biopsy of breast cancer dangerous to the patient? Cancer, 7:774, 1954.

88. Rosemond, G. P., Burnett, W. E., Caswell, H. T., and McAleer, D. J.: Aspiration of breast cysts as a diagnostic and therapeutic measure. Arch. Surg., 71:223, 1955.

89. Rosemond, G. P., Maier, W. P., and Brobyn, T. J.: Needle aspiration of breast cysts. Surg. Gynecol. Obstet., 128:351, 1969.

90. Rosen, P., Hajdu, S. I., Robbins, G., and Foote, F. W.: Diagnosis of carcinoma of the breast by aspiration biopsy. Surg. Gynecol. Obstet., 134:837, 1972.

91. Saphir, O.: Early diagnosis of breast lesions. JAMA, 150:859, 1952.

92. Schmidt-Matthiesen, H., and Glätzner, H.: Die Diagnostik des Mammacarcinoms. Frauenarzt, 16:192, 1975.

93. Schöndorf, H., and Naujoks, H.: Die Aspirationszytologie bei Mammatumoren. Fortschr. Med., 34:1400, 1974.

94. Schöndorf, H.: Die Bedeutung der Punktionszytologie bei der Abklärung verdächtiger Befunde in der Mamma. Therapiewoche, 26:812, 1976.

95. Scholz, A., and Scholz, C.: Brustdrüsenkrebs und Schwangerschaft. Zentralbl. Gynaekol., 93:10, 1971.

96. Schour, L., and Chu, E. W.: Fine needle aspiration in the management of patients with neoplastic disease. Acta Cytol., 18:472, 1974.

97. Smith, I. H., Fisher, J. H., Lott, J. S., and Thomson, D. H.: The cytological diagnosis of solid tumours by small needle aspiration and its influence on cancer clinic practice. Can. Med. Assoc. J., 80:855, 1959.

98. Söderström, N.: Fine Needle Aspiration Biopsy. Stockholm, Almqvist and Wiksell, 1966.

99. Soost, H., and Ries, P.: Die Zytologie der Brustdrüsensekrete und ihre Bedeutung für die Fruherkennung des Mammacarcinoms. Geburtshilfe Frauenheilkd., 28:918, 1968.

100. Stavrić, G. D., Tevcev, D. T., Kaftandjiev, D. R., and Novak, J. J.: Aspiration biopsy cytologic method in diagnosis of breast lesions. Acta Cytol., 17:188, 1973.

101. Stegner, H. E.: Pathologisch-anatomische Aspekte der organerhaltenden (konservierenden) Therapie bei Karzinomfrühstadien der Mamma. Österr. Z. Onkol., 2:136, 1975.

102. Tabár, L., Márton, Z., and Kádas, I.: Die Pneumoncystographie der Brust. Chirurg., 44:428, 1973.

103. Takahashi, M.: Color Atlas of Cancer Cytology. Tokyo, Igaku Shoiu, 1971.

104. Témime, R.: Contribution de la cytologie au diagnostic et à l'étude des tumeurs du sein. Bull. Fed. Soc. Gynecol. Obstet. Lang. Fr., 26:230, 1956.

105. Tischendorf, W.: Zytodiagnostik des Lymphknotenpunktats. Ergeb. Inn. Med. Kinderheilkd., 2:183, 1951.

106. Tribe, C. R.: Cytological diagnosis of breast tumors by the imprint method. J. Clin. Pathol., 18:31, 1965.

107. Vassilakos, P.: Tuberculosis of the breast: Cytologic findings with fine-needle aspiration. Acta Cytol., 17:160, 1973.

108. Verhaeghe, M., Cornillot, M., Herbeau, J., Wurtz, A., and Verhaeghe, G.: Le triple diagnostic cyto-radio-clinique dans les tumeurs du sein. Mém. Acad. Chir., 95:48, 1969.

109. Verhagen, A.: Tumor and Gravidität. Berlin, Springer, 1974.

110. Vilaplana, V. E., and Ayala, M. J.: The cytologic diagnosis of breast lesions. Acta Cytol., *19*:519, 1975.

111. Ward, G. R.: Bedside Haematology. Philadelphia, W. B. Saunders, 1912.
112. Webb, J.: The diagnostic cytology of breast carcinoma. Br. J. Surg., *57*:259, 1970.
113. Winship, T.: Aspiration biopsy of breast cancers by the pathologist. Am. J. Clin. Pathol., *52*:438, 1969.
114. Wurms, U., Schöndorf, H.: Zur Anwendung der Aspirationszytologie bei Rezidivtumoren des Mammacarcinoms. Arch. Gynaekol., *219*:142, 1975.

115. Zajdela, A.: La place du diagnostic cytologique par ponction dans les tumeurs du sein. J. Radiol. Electrol. Med. Nucl., *48*:682, 1967.
116. Zajdela, A., Ghossein, N. A., Pilleron, J. P., and Ennuyer, A.: The value of aspiration cytology in the diagnosis of breast cancer: Experience at the Fondation Curie. Cancer, *35*:499, 1975.

117. Zajicek, J.: Sampling of cells from human tumours by aspiration biopsy for diagnosis and research. Eur. J. Cancer, *1*:253, 1965.
118. Zajicek, J., Franzén, S., Jakobsson, P., Rubio, C., and Unsgaard, B.: Aspiration biopsy of mammary tumours in diagnosis and research: A critical review of 2200 cases. Acta Cytol., *11*:169, 1967.
119. Zajicek, J.: Zytologische Untersuchung von Punktaten in der Diagnostik der Brustdrüse. Schweiz. Med. Wochenschr., *99*:1271, 1969.
120. Zajicek, J.: Punktionszytologische Diagnostik von Veränderungen in der Brustdrüse. Wien Klin. Wochenschr., *82*:603, 1970.
121. Zajicek, J., Caspersson, T., Jakobsson, P., Kudynowski, J., Linsk, J., and Us-Krasovec, M.: Cytologic diagnosis of mammary tumours from aspiration biopsy smears. Acta Cytol., *14*:370, 1970.
122. Zajicek, J.: Aspiration Biopsy Cytology. Part I. Cytology of Supradiaphragmatic Organs. Basel, Karger, 1974.
123. Zinser, H. K.: Mammacarcinom, Diagnose und Differentialdiagnose. Stuttgart, Thieme, 1972.

Index

Page numbers in *italics* indicate illustrations; those followed by (t) indicate tables.